MOLECULAR APPROACHES TO ECOLOGY

This unique monograph stresses the importance of the comparative bio-chemical approach to the interpreta-tion of the adaptation of animals to their environment. It is based upon the premise that adaptation at the molecular level may not necessarily be recognizable by a study of single iso-lated molecules, but may result rather from changes at the level of several molecular species involved in the com-plex polygenic adaptive mechanisms. These mechanisms, from the morpho-logical down to the molecular level, are described in relation to their role in the colonization of the various en-vironmental media. In addition, the authors pose a number of suggestions concerning the molecular changes in-volved in the formation and evolution of animal species. In essense, the work provides the results of studies cen-tered around the systematic analysis of the molecular aspects subserving an adaptation recognizable at a higher level of organization.

Directed primarily to biologists, eco-logists, zoologists, and biochemists, this book will also appeal to the gen-eral reader with a fair knowledge of the fundamentals of biochemistry.

Molecular Approaches to Ecology

Molecular Approaches to Ecology

MARCEL FLORKIN AND ERNEST SCHOFFENIELS

Department of Biochemistry
University of Liège
Liège, Belgium

ACADEMIC PRESS   New York and London   1969

ACADEMIC PRESS, INC.
111 Fifth Avenue, New York, New York 10003

United Kingdom Edition published by
ACADEMIC PRESS, INC. (LONDON) LTD.
Berkeley Square House, London W.1

LIBRARY OF CONGRESS CATALOG CARD NUMBER: 68-28899

PRINTED IN THE UNITED STATES OF AMERICA

*La généralité doit ressortir
du fond même de l'analyse*

Claude Bernard
Cahier de notes (p. 90)

Contents

vii

Introduction

A century after the publication of Charles Darwin's great work, the theory of evolution is generally accepted by biologists. The filiation of organisms has wisely been derived mainly from morphological criteria. By the use of these criteria it has been possible to recognize changes at the level of molecular units (Florkin, 1944, 1966).[*]

Obviously, it is best to study those molecules endowed with a maximum of information. Unfortunately, however, it is more difficult to recognize the nature of the sequence of the four bases in DNA and RNA than the sequence of the twenty amino acid units which form a protein. Variations at the level of the primary structure of proteins—considered "along the branches of the phylogenic tree"—have been a favorite subject of study for the biochemist. By studying such changes as occur phylogenically in cytochrome c, hemoglobin chains, insulin, pituitary hormones, fibrinogen, trypsinogen, chymotrypsinogen, etc., it has been revealed that homologous proteins rarely show a change in the number of amino acid residues, but present substitutions, generally bearing—as in the case of hemoglobin—on a single amino acid. Most proteins in an individual are identical to those of other individuals of the same species. There are, nevertheless, individual variations, and the case of hemoglobin, for example, shows that individuals carrying an abnormal protein are apt to be found in limited geographic areas. This suggests that amino acid substitutions appear sporadically. How are they eventually extended to become included in the genome of the whole species? The classic answer is natural selection. But, how can the substitution of aspartic acid in position 62 of cytochrome c in man for

[*]Florkin, M. (1944). "L'Evolution Biochimique." Masson, Paris.
Florkin, M. (1966). "A Molecular Approach to Phylogeny." Elsevier, Amsterdam.

glutamic acid in the horse, for example — a change which implies no modification in the activity of the molecule — be linked to natural selection? While increasing our knowledge of the variability of polypeptide chains in phylogeny, and consequently of the corresponding DNA chains, such data bring no information concerning the process of evolution at the molecular level. They do not illustrate the innovative aspects of natural selection, but rather its conservative effects. In research on protein structure and protein degrees of homology, biochemists usually start by comparing an activity such as that shared by all hemoglobins, by all cytochromes c, and by all insulins. They find the uniformity of the structures underlying these activities accompanied by trivial substitutions. Consequently they comprehend only the conservative effects of natural selection.

Innovations in evolution resulting from natural selection are of an adaptive nature. Although the concept of adaptation at the molecular level was stated some time ago (Florkin, 1944), it has only been possible to identify it in exceptionally favorable cases.

It has for many years been our conviction that adaptation at the molecular level may not necessarily be recognizable by a study of single isolated molecules. Rather, such adaptation may result from changes at the level of several molecular species involved in the complex polygenic adaptive mechanisms. A systematic analysis of the molecular aspects underlying an adaptation recognizable at a higher level of organization has become a favorite subject for study in our laboratory. Some of these studies are reported in this work, along with discussion of the concept of molecular adaptation.

Our warmest thanks are due to our colleagues Professor C. Jeuniaux, Dr. G. Duchateau-Bosson, and Dr. S. Bricteux-Gregoire, who have contributed their share to the progress of this austere trail of research, and to Mrs. A. Closset, for typing the manuscript and helping to prepare the bibliographic material.

November, 1968   MARCEL FLORKIN
                                                                              E. SCHOFFENIELS

CHAPTER I

Adaptation and Natural Selection

Organisms are made of cells, of modified cells, and of products of cell activities. Each cell is a constellation of macro-molecules, molecules, and ions assembled according to a common scheme, also prevailing in unicellular organisms. Not only the architecture of this scheme, but also the nature of its constituents show a certain degree of similarity from cell to cell. This similarity is the reality behind the concept of the "unity of life." The "unity of biochemical plan" is nothing more than the cell theory stated in plain chemical terms. As already proposed by the founder of the cell theory, Theodor Schwann, in his epoch-making little book published in 1839, cells are *units of metabolism* in the same way as they are *units of structure*. The unity of structure and metabolism in the whole of living beings is an expression of cellular continuity and of the persistance, through this continuity, of a collection of definite sequences of purine and pyrimidine bases, which are the sequences that control the biosynthesis of the collection of enzymes found in each cell. In spite of certain degree of unity in the nature of at least a part of this collection, which is only the canvas on which the cells have embroidered the diversity of their differentiation, no cell is limited to the underlying chemical similarity; the "unity of plan" remains an abstraction in the human mind.

The increasing interest found in the molecular approach to biology induces us to consider a species as consisting of groups of individuals having closely related combinations of purine and pyrimidine bases in their macromolecules of deoxyribonucleic

1

acid (DNA), and of which the systems of operators and repressors bring about the biosynthesis of similar sequences of amino acids. Their integration in the cell leads to similar structural and functional characteristics, adapted to the ecological niche in which the species succeeds.

Evolution may be defined as the changes that affect the relative proportions, the associations, and the nature of genetic factors within a population. Speciation is the means through which new aspects of evolution are introduced; it is worked upon by the nature of the environments and what is generally called adaptation results from this influence. The environment is one of the agents of natural selection that produces its effects by affecting the rate of reproduction. A chemical modification of genetic factors is the most complete achievement of the molecular aspect of evolution. Ontogeny and phylogeny thus result form the reading of the code transmitted by the genotype, while evolution is a consequence of the existence of new codes or of new decipherment of the code.

Since Lamarck, we have learned to consider the diversity of species as being the result of adaptations to different environmental conditions. Organisms have responded to the solicitations of the surroundings either by the diversification of the genotypes or by an adjustment of the translations of the code to the new conditions.

Adaptation, as understood here so far, is an organismic or evolutionary concept. It explains why a species survives in or is able to colonize a given environment. In a more teleological sense one could also say that adaptation functions for the survival of the species (Williams, 1966).

Within the complex picture of all the adaptive features it is possible to individualize biochemical traits: one thus defines the biochemical adaptations: i.e., the molecule(s) or molecular mechanism(s) that explains, if not all, at least part of the evolutionary adaptation.

If for example, we consider the oxygen absorption curves of the few species for which we know the respiratory cycle (Fig. 1-1), we see that in the different cases, even if the external medium is the same, very different values of the gradient of oxygen pressures obtain between the environment and the milieu intér-

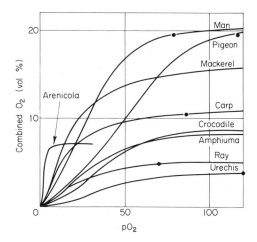

FIG. 1-1. Oxygen absorption curves for various bloods and coelomic fluids containing hemoglobin under "arterial" conditions reproduced *in vitro*. The black circle indicates the value of the arterial $+O_2$ (Florkin, 1960).

ieur. The partial pressure of oxygen corresponding to "arterial" conditions is obviously related to the rate of exchange through the respiratory epithelium, gills or lungs as the case may be, as well as to that of the circulation of the oxygen carrier. In spite of this, the degree of saturation of the arterial blood corresponds in each case to the higher part of the dissociation curve. As a matter of consequence, any lowering of pO_2 in the medium results in the delivery of oxygen from the carrier (Florkin, 1948). Such a biochemical adaptation is evidently only one among many features that explains the relationship a given species establishes with its surroundings and it is thus of prime importance to define the hierarchy of adaptations not only at the organismic or evolutionary level but also at the molecular scale.

A proper approach to the study of adaptation must start from the consideration of the relation organism-environment at the level of the community or of the organism, and proceed progressively from this organismic starting point to the underlying molecular aspects. Such a study, to be fully convincing, must be very detailed. It has been accomplished in the case of the adaptation of euryhaline invertebrates (Chapter VII). In certain of these euryhaline forms, the adaptation results from the combination of the effects of an anisosmotic regulation of the body fluids kept at a higher concentration than the external medium, and of an isosmotic intracellular regulation bringing the cells into osmotic equilibrium with the body fluids.

Isosmotic intracellular regulation results from several biochemical adaptations, one of which, the regulation of the total concentration of free amino acids and taurine, is always playing an important role in the adaptation. The analysis of this regulation has led to its location at the level of a differential action of inorganic cell constituents on several enzymes related to the amino acid metabolism and to the production of reducing equivalents. In this case, the evolutionary adaptation is partly explained by the peculiar properties of some protein molecules. The biochemical adaptation thus lies in certain features of the enzyme that place its catalytic properties under the control of the inorganic constituents of the cell.

In other words this adaptation is at the level of the protein structure. Consequently one is tempted to establish a direct relationship with the genotype.

The significant conclusion to be drawn from the study of a metabolic pathway throughout the animal kingdom or, when considering the same species, in various organs, is that an important aspect of adaptation rests not only on the existence of peculiar catalytic steps leading to specific metabolic products, but also on the development of specific means of control of a catalytic step common otherwise to any type of cell. This control, generally allosteric in nature, explains how the activity of a metabolic route may have various functional significances according to the differentiation considered. This conclusion stemming from the careful investigations of research workers involved in the comparative study of enzyme systems has an important consequence. It indeed focuses attention on the problem of the structure of the enzymes catalyzing the same reaction in a large diversity of cells. If according to the origin of the cell the enzyme is variously affected by controlling factors, this suggests that the differentiation or the adaptation, as the case may be, deals primarily with the development of control sites separated from the catalytic centers. Available data suggest that for a number of enzymes the genetic information responsible for the amino acid sequence determines automatically the secondary and tertiary protein structure. The assumption that a folding template carries information differing from that contained in the amino acid code has not yet been supported by experimental findings. We may thus expect that the emergence of allosteric sites is related to the primary structure of the protein. The possibility that cytoplasmic environmental influences may, in a given cell, affect the protein configuration that is responsible for the appearance of allosteric sites has still to be investigated. Another factor that may modify profoundly the biological function of an enzyme is its intracellular location. All remains to be accomplished as to the factors that may induce this distribution.

Another evolutionary adaptation studied in great detail in this book deals with the spinning of a cocoon by the silkworm. The physical properties of fibroin, underlining the protective function of the cocoon, depend on the peculiarities of protein synthesis in the distal part of the silk gland. The structure of fibroin may be explained, according to current theories, by the specific nature of the mRNA liberated therein. But the detailed analysis shows that if the explanation of the *composition and structure* of fibroin is at the level of a gene, the *existence* of the cocoon depends on other factors as well: the nature of the food of the silkworm, its attraction by mulberry leaves and the aspects of behavior leading it to masticate and to swallow the leaves, the enormous appetite of the silkworm during the second part of the fifth instar, the characteristics of the substraction of the amino acids from hemolymph, the regulation of the nymphal weight, etc. All these factors contribute to the very existence of a cocoon and are revealed by the analysis of the biochemical adaptations which clearly are polygenic.

Rather paradoxically, the concept of "molecular biology" is really an organismic concept. The authority of its inventors forces us to accept this view. If we read the Instructions to Authors of the *Journal of Molecular Biology*, we learn that molecular biology is the domain of studies related to the nature, production, and replication of *biological structures considered at the molecular level* and to the relations of these structures with the *functions of organisms*. It appears clearly, at least in the current theories, that heredity (a function of organism) is linked to the nature of certain sequences of purine and pyrimidine bases within the macromolecules of DNA of the genes (i.e., to a biological structure considered at the molecular level).

If the animals are generally endowed with motility (function of organism), it is related to the fact that certain cells biosynthesize macromolecules of contractile proteins (molecular structures related to the function considered at the level of the organism). In fact, molecular biology, in spite of the situation of its studies at the molecular level, brings us back to the organismic viewpoint too often neglected in biochemical studies.

This attractive and fruitful tendency should, nevertheless, not be considered as being universally applicable. Some macromolecules are active, not at the level of the organism, but at another level of organization. We hope to be able, at some time, to climb the staircase of causal relations through the gradation — molecule, cellular organelle, cell, tissue, organ, individual, population, species, community, ecosystem — but this time has not yet arrived. The macromolecules studied at the structural level by orthodox molecular biologists are those whose function are kept through all levels of organization and that remain unchanged at the level of the organism as a whole. There are exceptions and there are many cases of different functions assumed at the level of the organism by the same macromolecules (physiological radiations). This does not tend to contradict the primary interest of biological structures considered at the molecular level and of their relations with whatever level of organizations they contribute to.

On the other hand, the authors do not intend to deter the deep interest and importance of population dynamics. It is certainly true, as Grant (1963) has pointed out, that the presence of coyotes is a factor in the determination of the number of rabbits living in a given territory. As Grant rightly states, an increase in the population of coyotes will reduce the rabbit population. The expansion of the population of coyotes will eventually stop when the number of rabbits will have reached the limits compatible with the number of rabbits they eat. No molecular approach will reveal these causal relations. Everybody will agree to this. But whatever patience and ingenuity may be devoted to counting trouts and insects living in a stream, it is only by a molecular approach and through the knowledge of the properties of trout hemoglobin that we shall understand why trouts live in streams and not in marshes, where insects also exist.

It is at the level of the organism that the impact of natural selection takes place. In the case of an evolutionary adaptation allowing a species to live in the brackish water of estuaries as well as in seawater, the advantage conferred by the adaptation at the molecular scale of dimension is obvious, as it allows the colonization of new biotopes. It is this aspect and its nutritional consequences which give rise to an increase in the number of individuals reaching the age of reproduction and the number of their offspring. This is the level where the phenomenological aspect of the evolution obtains, while the causal factors are located at the molecular level of protein structures.

All these considerations show that in the study of adaptations it would certainly be ill-advised to dissociate the molecular approach from its organismic counterpart.

References

Florkin, M. (1948). *Experientia* 4, 176.

Florkin, M. (1960). "Unity and Diversity in Biochemistry." Pergamon, London.

Grant, V. (1963). "The origin of adaptations." Columbia Univ. Press, New York.

Schwann, T. (1839). "Mikroskopische Untersuchungen über die Uebere-instim-
mung in der Struktur und dem Wachstum der Thiere und Pflanzen." San-
derschen Buchhandl., Berlin.

Williams, G. C. (1966). "Adaptation and Natural Selection." Princeton Univ.
Press, Princeton, N.J.

CHAPTER II

Basic Concepts of Comparative Biochemistry

To avoid misunderstandings, we should agree at the start on the definition of some of the basic concepts of comparative biochemistry. The biochemical compounds, molecules or macromolecules, which show signs of chemical kinship, we shall call *isologs*. Cytochrome, peroxidase, catalase, hemoglobin, and chlorocruorin show this isology, as they are heme derivatives. In the case of the hemoglobins of two men who are identical twins, the maximum degree of isology obtains. It is less pronounced when we consider the hemoglobins of a dog and of a jackal, and still less if we consider those of a dog and of a horse. In these cases, the protohemes are identical, and the degree of isology depends on the structure of globin, the protein moiety of the hemoglobin macromolecule. If we compare a hemoglobin and a cytochrome c, the isology obtains only at the level of the heme moiety, the sequences of amino acids in the protein part being nonisologous (Margoliash and Tuppy, 1960). Isology is a chemical concept.

Extensive studies on the primary structure of a number of macromolecules of proteins (see Vegotsky and Fox, 1962) and the knowledge that this primary structure is a reflection of the sequences of bases of DNA indirectly controlling its biosynthesis have led to the conclusion, for instance, that the "insulin-determining" sequences of bases of the pig and of the sperm whale are identical. From studies on ACTH, ribonuclease, melanotropic hormone, insulin, cytochrome c, hemoglobin, etc., it appears that if each of these macromolecules is obtained from

8

different organisms, extensive similarities in primary structure are observed, as Fig. 2-1 shows in the case of cytochrome c. This degree of isology is incompatible with chance effects and points to a permanence, through the whole of the evolutionary tree, of very ancient base sequences of DNA. These sequences, reproduced through the ages with more or less constancy, may be called *homologs* in the sense used by the biologist, i.e., connoting a common origin and a common ancestry in descent. *Homology* is thus, at the molecular level, considered to be not a chemical concept but a genetic concept, which differs from the use of the word by enzymologists, where it means a similar function. In the concept of homology as presented here, the function and the action are irrelevant, and the concept is one of common origin by replication of a base sequence of DNA from a common ancestor. The whole of our present knowledge brings us to accept, with the usual scientific wariness, that there is a great probability that the very isologous primary structures of proteins are patterned after very isologous base sequences in nucleic acids and consequently that they also can be qualified as homologous (direct homology). Homology can be extended to sequences of homologous protein biocatalysts, and also to the results of a biosynthesis catalyzed by a homologous enzyme chain (indirect homology). This definition makes clear the distinction between isology and homology. ATP is isologous in all cells, but it is not always ho-

┌Heme┐
–Gly– – –Lys.Gly– – –Phe– – –CyS– –CyS.His.Thr.Val.Glu–
                10   20
Gly.Gly–His.Lys–Gly.Pro.Asn.Leu–Gly–Phe.Gly.Arg– –Gly.Gln.Ala–
                   30   40
Gly– –Tyr.Thr–Ala.Asn– –Lys– – –Try–Glu– – – – –Tyr.
        50   60
Leu–Asn.Pro.Lys.Lys.Tyr.Ileu.Pro.Gly.Thr.Lys.Met.Ileu.Phe–Gly–
    70   80
Lys.Lys– – –Arg–Asp.Leu– –Tyr.Leu.Lys.Lys– – – –COOH
      90   100

FIG. 2-1. Identical residues in the amino acid sequences of all actually known vertebrates and yeast cytochromes c. Numbering of the residue positions follows that of the vertebrate proteins; yeast cytochrome c has five extra residues at the N-terminal and one less residue at the carboxyl terminal. The identical residues in all species are given and the 58 nonidentical residues are indicated by a dash. (Smith and Margoliash, 1964.)

mologous, since, for instance, the product of the action of a chain
of biocatalysts in glycolysis and of another in oxidative phosphorylation. Bile acids, on the other hand, are homologous in
all vertebrates, as they are synthesized in pathways catalyzed by
homologous enzyme chains.

The term *analogous* is applied to biochemical units playing a
similar role in biochemical systems. The luciferins are analogs,
though they are not isologs. Another most interesting example of
analogy is provided by the different kinds of oxygen carrier:
hemoglobins, chlorocruorins, hemocyanines, and hemerythrins.

Phosphagens are another case in point. This name has been
given by Eggleton and Eggleton (1927) to a category of phosphorylated derivatives isolated from muscles. They are all N'-
phosphorylguanidine compounds bisubstituted as in the case of
phosphocreatine or monosubstituted as in the case of phosphoarginine, phosphoglycocyamine, phosphotaurocyamine, phospho-
hypotaurocyamine, phospholombricine, and phosphoopheline.
They are labile and easily hydrolyzed at 37°C in slightly acid
solutions. The free energy of hydrolysis amounts to about 10 kcal
$Mole^{-1}$. Their role in the cell is to control the content in ATP
through the reversible functioning of the so-called Lohman
reaction:

$$ATP + creatine \rightleftharpoons phosphocreatine + ADP$$

As such, phosphagens are analogous compounds.

By studying the distribution of phosphagens among the animals some workers have been led to propose that phosphocreatine would only be present among the vertebrates while phosphoarginine would characterize the invertebrates. It has even
been suggested that the simultaneous presence of both phosphagens in a given group would point to the ancestor of the vertebrates. This theory has had the destiny it deserved by over-
looking two essential methodological exigences (Florkin, 1966).
First, any phylogenic consideration as applied to biochemical
constituents must rest on the phylogeny of organisms as derived
from the data patiently accumulated by generations of naturalists. Second, homologous molecules are the only chemicals for
which phylogenic relations may be sought. If it is true that the
known function of all phosphagens is to buffer the cell concen-

tration in ATP, their chemical structure is directly related to the
system responsible for the biogenesis of guanidine bases as well
as to the specificity of the transferases (Thoai and Roche,
1964a,b).

Analogy, as we have said, refers to nonisologous structures
presenting similar activities. On the other hand, if a common
primary structure, a sequence of amino acids, were discovered in
different cells where it could be proved that the initial prototypes of these sequences were different (a most improbable
event according to our present theories), we ought to consider
this as a convergence. When a same biochemical compound
appears in different cells as the result of the action of different
enzymatic systems, we shall speak of parallelism.

The concepts of isology, homology, analogy, and convergence
are particularly clear when applied to the consideration of primary protein structures, i.e., of amino acid sequences and to
their initial prototypes made of sequences of purine and pyrimidine bases in nucleic acids (Fig. 2-2).

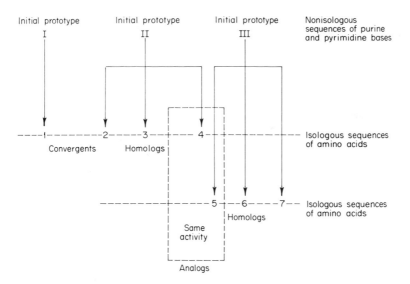

FIG. 2-2.   Isology, homology, analogy, and convergence. The roman figures
designate sequences of purine and pyrimidine bases. The arabic figures indicate
sequences of amino acids. (Florkin, 1962.)

Molecular evolution takes on a number of forms in the history of living organisms. It can take the form of a change in structure of the molecule or macromolecule, in descent. The known aspects of molecular changes along the lines of the phylogeny of multicellular organisms have recently been reviewed (Florkin, 1966). We are concerned with the phylogeny of a protein when we establish comparisons between homologous forms of this protein at different levels of the same phyletic series of the organisms synthesizing these forms. The beautiful studies of Ingram and his colleagues on the evolution of hemoglobin provides us with a good example. It has been clearly emphasized that this has a great ecological impact. Data of this kind have also been collected for the phylogeny of fibrinopeptides, of the peptide liberated by the activation of trypsinogen, of the neurohypophyseal hormones, etc.

Other forms of molecular phylogeny can be detected among molecules that result from the extensions of a biosynthetic pathway. In that case the various steps leading to the final product repeat its phylogeny. Molecular changes in descent can often be related, in these cases, to adaptive aspects. More and more examples will certainly be found as our knowledge of the biochemical continuum increases.

But are the molecular changes that are detectable at the level of primary structure along the paths of phylogeny[1] functional and adaptive? This certainly does not appear to be the case when considering the phylogeny of the protein so far studied. Nevertheless as stated by Simpson (1964), "it is certainly not true as a generalization that molecular differences among species are commonly nonfunctional or inadaptive." In the field of indirect homology, molecular differences are clearly adaptive in many cases. Considering the biochemical continuum (Chapter III) we meet problems of the biosynthesis of new molecules clearly of an adaptive nature, the integrated action of which is linked to a specific action on the receptor organism.

In direct homology, our ignorance of the possible adaptive nature of the changes appearing in the primary structure may be

[1]Bacteria are left out of the picture as their phylogeny is derived from molecular aspects.

the only reason that leads us to conclude that the change is not adaptive. Therefore instead of reasoning from the molecular up to the organismic aspect, the problem must, at least tentatively, be tackled the other way round: adaptations clearly defined at the organismic level should be analyzed at the molecular level of dimension. This is done in Chapter VII.

A change of primary structure in a protein chain always depends on a change of sequence in a nucleic acid. While the change of structure in homologous nucleic acids, and consequently of proteins, is the leading thread of the phylogeny of macromolecules, this form of change is not the only method of molecular evolution. Even if it has kept, more or less modified, some characteristics of an ancient prototype in parts of its amino acid sequences, a protein may acquire new kinds of activities and properties by changes in its secondary or tertiary structures. Therefore, the primary structures, at least in the cases of marked isology, will give us information concerning molecular phylogeny while the truly adaptive property will reside in the configurational change. The exact contribution of the amino acid sequence and of possible cytoplasmic factors in determining the tertiary structure of the protein has still to be defined (see Chapter I, p. 5).

The guide of molecular phylogeny remains, at least for the time being, the phylogeny of organisms as built by generations of naturalists. Evolution, with change of structure, has probably been most active at the level of the precellular eobionts, when the architecture we now find as common to all cells had been acquired by natural selection in the first reproducing organisms. But since then, this kind of molecular evolution has continued to take part in the ecobiochemical development of relationships between organisms and the medium during the expansion of the biosphere. Change of structure is eventually accompanied by a change of action.

Other forms of molecular evolution have played an ecological role. Systems of macromolecules can be modified by the change of structure of one or several of their components, by modifications in the relative concentrations of these components, by the loss of one or several of these, by the introduction of an existing macromolecule into another system. One of the aspects of the molecular evolution in vertebrates is the acquisition, at the level of the mesodermal cells, of new steroid biosynthesis. One of the consequences of this is ionic regulation by the action of corticosteroids at the level of the nephridia. This action is due to the introduction of the corticosteroids in the systems of transport at the level of the urinary tubules. A new form of molecular evolution, taking place in amphibians, consists of the introduction of pituitary hormones in the system, with the consequence that water is reabsorbed — an important aspect of the factors allowing for terrestrial life in the toad, for instance.

On the other hand, a biochemical system, whether it remains the same or undergoes a change of structure and eventually of *action,* may undergo a change of function when considered at a level higher than molecular. This results from a different insertion in the tridimensional system of the cells, and constitutes another form of adaptive radiation.

The system of the biosynthesis of thyroid hormones appears in Tunicata and its course is at the start still the same as it remains in vertebrates, although no particular organ is differentiated for this function in tunicates, and although the role played by the thyroid hormones in these animals has not yet been identified. In Cephalochordata, such as *Branchiostoma (Amphioxus lanceolatum)*, iodine concentrates in a special organ, the endostyle. Here, the secretion takes on the holocrine mode, the endostyle cells in which the hormones are synthesized going toward the digestive tube where they are hydrolyzed and where the thyroid hormones are recovered by intestinal absorption. In larval cyclostomes, the same mode prevails, but when they reach the adult stage, the thyroid hormones are liberated by merocrine secretion oriented toward the circulating blood. This physiological radiation brings the thyroxine and the triiodotyrosine to the fulfillment of the important function they play in the higher vertebrates (Roche, 1963).

The change of function of an enzyme system such as the chitinolytic one, formed by chitinase and chitobiase, shows a number of ecological aspects, its effect having the role of entering in digestion, or in molting, or in the penetration of parasites in their host (Jeuniaux, 1963).

From what has been stated above, it appears that different aspects of molecular evolution have been used in taking advantage of environmental fitness. They can all provide ecobiochemical viewpoints, whether or not they are of taxonomic or of phylogenic importance. The examples of analogous substances, as well as evolution with a change of molecular structure, are ways of ensuring the diversity of properties of biochemical components playing similar roles. This diversity is one of the conditions of the integration and of the regulation of the organism and of its integration into the ecosystem. In the history of comparative physiology, homologous and analogous components have not been distinguished. In our present consideration, this fundamental distinction must be present in the first rank, for the benefit of a sounder appreciation of the adaptive nature of molecular evolution and of the site of impact of natural selection.

References

Eggleton, P., and Eggleton, G. P. (1927). *Biochem. J.* **21**, 190.

Florkin, M. (1962). *Bull. Classe Sci., Acad. Roy. Belg.* **48**, 819.

Florkin, M. (1966). "Aspects moléculaires de l'adaptation et de la phylogénie." Masson, Paris.

Jeuniaux, Ch. (1963). "Chitine et chitinolyse, un chapitre de la Biologie moléculaire." Masson, Paris.

Margoliash, E., and Tuppy, H. (1960). Communication presented at the 138th annual meeting of the American Chemical Society, New York. (Cited by Zuckerkandl and Pauling, 1962).

Roche, J. (1963). *In* "Evolutionary Biochemistry" (A. I. Oparin, ed.) pp. 313–326. Pergamon, Oxford.

Simpson, G. G. (1964). *Science* **146**, 1535.

Smith, E. L., and Margoliash, E. (1964). *Federation Proc.* **23**, 1243.

Thoai, N. von, and Roche, J. (1964a). *In* "Taxonomic Biochemistry and Serology" (C. A. Leone, ed.) pp. 347–362. Ronald, New York.

Thoai, N. von, and Roche, J. (1964b). *Biol. Rev. Cambridge Phil. Soc.* **39**, 214.

Vegotsky, A., and Fox, S. W. (1962). *In* "Comparative Biochemistry" (M. Florkin and H. S. Mason, eds.) Vol. IV, pp. 185–244. Academic Press, New York.

Zuckerkandl, E., and Pauling, L. (1962). *In* "Horizons in Biochemistry" (M. Kasha and B. Pullman, eds.) pp. 189–225. Academic Press, New York.

CHAPTER III

The Biochemical Continuum and Ecological Integration

The concept of the biosphere defined as representing the portion of the earth and its atmosphere that is capable of supporting life is useful in the field of biogeochemistry. From the ecological viewpoint it is also useful to consider a *biochemical continuum* of the total of the living mass and its metabolic extensions. The oxygen of the atmosphere has been produced by the living plants and the CO_2 it contains has a clear relationship with the metabolism of organisms. The sediments at the deepest bottoms of the oceans, and the organic matters coating sand or mud particles there, are linked to the surface of the seas by a cloud of molecules—this cloud being of variable density. The same notion applies to soils, to fresh waters, and even to sedimentary rocks in which a part of the organic matter of sediments persists in the form of kerogen. The whole of the biochemical continuum forms a population of molecules densely associated in parts and more separated in other regions. Some of the dense parts are architectured in the form of living cells. At this level of organization, the population of molecules is kept away from the state of equilibrium, and a mechanism has been developed for reproduction and information transfer. Monocellular and pluricellular organisms associate in communities whose populations maintain themselves and form, in association with different components of the environment, ecosystems, inside which and between which, currents of matter and energy take place.

The biochemical continuum is the result of a slow evolution

starting from an organic abiogenic continuum. According to current ideas the cosmos originated in the explosion, five billion years ago, of a dense core of neutrons. The chemical elements appeared very rapidly. Most of the cosmic matter is dispersed among galaxies and stars as atoms, molecules, or cosmic particles. Hydrogen is the most abundant constituent and CH molecules are readily formed. The interstellar space also contains CN molecules. These molecules as well as C_2 are present in the atmosphere of relatively cool stars. The head and tail of comets contain a series of radicals such as C_2, C_3, CN, CH, OH, CO, and N_2.

In the cloud of gas and cosmic dust that surrounded the sun before giving rise to the planets of the solar system, the simple carbon compounds found in the interstellar spaces were certainly present and, according to the astrophysicists, they did participate in the formation of the earth.

Many organic compounds have been identified in meteorites (see Hayes, 1967), thus confirming the reality and the universality of an organic cosmochemistry. Organic chemistry, i.e., that of carbon derivatives, is as old as the world, and its numerous aspects are certainly extended through the entirety of the universe.

If we consider more specifically our planet, it is generally accepted that in its early stages of formation, the earth's atmosphere was composed of H_2, CH_4, NH_3, and H_2O. This is the present composition of the atmosphere surrounding planets, such as Saturn, situated far away from the sun. The reducing primitive atmosphere has been oxidized: N_2 has replaced NH_3, and, after the invention of photosynthesis, O_2 took the place of part of the water. Many organic compounds have also been formed from methane and ammonia during that period of intense oxidation. It is possible to reproduce experimentally these processes of abiogenic synthesis: many data have been collected showing that the action of electrical discharge or ultraviolet radiation on a mixture of methane, ammonia, water, and hydrogen give a significant yield of amino, hydroxy, and aliphatic acids (Miller, 1953, 1957).

Many other molecules among those entering into the composition of organisms have also been synthesized in this way: carbohydrates, porphyrins, purines, pyrimidines, and even small peptides (see Oro, 1963). We are thus allowed to consider that the chemistry of carbon derivatives, traditionally designed as organic chemistry, cannot be defined by the nature of the synthesis performed by the organisms, but that the organisms rather found their constituents in a cosmic chemistry, the effect of which has brought about the presence of organic molecules in the primitive ocean. Their concentration was such that many transformations were possible, synthesis and hydrolysis as well, through the interplay of reactions finding their energy in the solar radiations. Thus it is from this abiogenic organic continuum that the complex system of the biochemical continuum has evolved.

If one considers some properties of the elements which compose the bulk of the biochemical continuum (H, C, N, O, S, P), one sees that, except for hydrogen, they are those that readily form multiple bonds. Even when engaged in single bonds, they dispose of lone pairs of electrons that may be subjected to delocalization and help to bridge conjugated fragments. This has led to the formation of molecules the structure of which is that of conjugated systems with complementary properties of stability, e.g., toward ionizing radiations and reaction possibilities not encountered in other types of molecules. This makes them particularly well-suited to compose the array of substances taking part in the chemical reactions of which the cell is the site. As noticed by Pullman and Pullman (1962, 1963) the presence of mobile electrons is responsible for many of the unique properties with which the molecules of the biochemical continuum are endowed. It is certainly reasonable to assume that they were selected for that very reason in the course of the chemical evolution preceding the apparition of life.

As stated above, regions of the abiogenic continuum became organized to give the primitive cells. In its accomplished state, certainly not realized at once, the complex network of metabolic sequences that describes a cell is determined by a system whose structure is well understood, since the fundamental discovery of Watson and Crick (1953): the double helicoidal structure of deoxyribonucleic acid (DNA). The transition from an abiogenic continuum to a primitive cell implies the formation and the reproduction of polymers, e.g., proteins and nucleic acids, the con-

stituents of which were certainly present in the soup of the pri-
meval ocean.

As stated by Van Niel (1956), "Acceptance of the postulate that
chemical evolution preceded biopoesis further suggests that the
organized structures representing primitive life were capable of
self-reproduction before they acquired mechanisms by means of
which they could chemically transform the components of their
environments."

This property of self-organization and self-reproduction has
been the object of thorough investigations by Fox and his colleagues (see Fox and Yuyama, 1963; Fox et al., 1967). They have
shown that by heating mixtures of amino acids at 150 − 180° C, it
is possible to obtain polypeptides (proteinoid). If these polymers
are dissolved in a solution of NaCl and heated, they form micro-
spheres, spherical in shape in the range of 0.5 to 80μ in diameter.
They exhibit some properties of proteins including catalysis.
They also develop budlike appendages which, when separated
and transferred to fresh proteinoid solution, can grow in mass to
new particles. It should however be made clear that the micros-
pheres of Fox should not be considered as protocells, and Fox
himself has never stated that the first proteins were obtained in
this way. The results simply show that one may obtain a synthesis of proteins in conditions that differ from those actually
found in contemporary organisms. This being said, it is easy to
conceive the formation of a proteinaceous proliferating system in
the primitive ocean as well as polymers of nucleotides. These
have been obtained by Ponnamperuma (1965) by the action of
ultraviolet light on a mixture of purines, pyrimidines, sugars, and
phosphoric acid. More recently the synthesis in vitro of DNA has
been reported by Kornberg and his associates (see Singer, 1967)
which indicates, among other things, that no unusual linkages
occur over the many nucleotides forming a genome.

In relation with the above arguments is the question of
whether genic mechanism preceded or followed the apparition
of an organized structure with cell-like properties. Many views
have been expressed in this respect (see for instance: Calvin,
1962; Granick, 1965; Lipman, 1965; Fox et al., 1967).

One may suppose with Jukes (1966) that preenzymatic proteins have been incorporated in a spherical bileaflet boundary structure and that they have translated their amino acid sequence into ribonucleic acid through the intervention of a RNA-polymerasic protein. Such an RNA molecule could then be copied in a DNA structure following a method analogous to that used in the experiments of hybridation. The proteins could be synthesized according to the mechanism prevailing in contemporary cells. The code, as known today, is responsible for the translation of the sequence of purine and pyrimidine bases of the molecules of DNA in terms of proteins. These molecules have been subjected to the hazard of changes, thus leading to what is generally called molecular evolution. As a consequence the phenotype is modified and it is at this level that natural selection exerts its most profound influence. Therefore one could propose that the adaptation in pluricellular organisms is the result of complex relations between the multiple aspects of this molecular evolution. Accordingly, the nature of the adaptation could not be situated at this level. However, throughout this book many properties of molecules that are clearly adaptive are presented. Moreover, in the course of evolution of the organisms, relations that are clearly mediated by molecules have been established between them. Many examples may be found at the level of the biochemical continuum as shown below.

The transition from the abiogenic organic continuum to the biochemical continuum is characterized by an important step: the use of organic molecules by organized aggregates (precell or protocell). This is the origin of nutrition. After the invention of photosynthesis, an organized aggregate becomes a producer of organic molecules serving as "food" to the forms that were lacking the systems of photoautotrophy. Progressively, the reciprocal relations between the various systems became more and more elaborate, which led to the exchanges of molecules that were not only of nutritional importance but acquired also regulatory and control properties on the metabolism. Good cases in point are given by the consideration of the following examples.

In Iran and Egypt a disease that affects only human males is characterized by a delay in growth and a hypofunctioning of the testis. It is rapidly cured by the adjunction of zinc in a diet, otherwise poor in this metal, and which is unable to compensate the loss due to excessive sweating and to infection by hookworms. The deficiency in zinc manifests itself at the level of metallo-proteins endowed with catalytic properties such as the glutamic dehydrogenase. In chicken, a lack of copper in the food is asso-ciated with a decrease in the aminoxidase activity of the aorta. This results in a reduced capacity for oxidatively deaminating the ϵ-amino groups of the lysine residues in elastin. As a conse-quence less desmosine is formed. Since it is the cross-linkage group of elastin, fewer cross-linkages are formed in this protein corresponding to a marked decrease in the elasticity of the aorta (Hill *et al.*, 1967).

A deficiency of manganese is associated in the chicken with malformations in the legs due to a marked reduction of chondro-itin sulfate in the epiphyseal cartilage (Leach, 1967).

The literature shows that dissolved organic carbon amounts to between 1 and 5 mg per liter in the oceans. The sources of this organic carbon are found in the decomposing organisms, in the compounds secreted or excreted by them, in land sources, and in an exchange with sediments (Plunkett and Rakestraw, 1955).

A large number of organic molecules have been isolated from soils, lakes, and oceans (Vallentyne, 1957) and their ecological importance recognized in a number of cases. In 1937, for in-stance, at the Plymouth Marine Biological Laboratory, H. W. Harvey observed that diatoms did not show any satisfactory growth in the waters in the vicinity of Plymouth. He showed that some organic factors were missing. Many observations have confirmed, since that time, the need of the presence of growth factors in seawater in order to ensure a satisfactory production of phytoplankton. One of the factors governing algal bloom may be vitamin B_{12}.

"We know that cobalamins are present in the littoral and es-tuarine zone, a good proportion of marine bacteria are B_{12} pro-ducers, and several seaweeds actually absorb these compounds. With the circumstancial evidence at hand, we can postulate that the cobalamins might be present at a suboptimal level, support-ing endemic populations of B_{12}-requiring marine flagellates, and certain blooms may be induced by a sudden increase of B_{12} in the water. In the littoral zone this may coincide with the influx of fresh water from rivers and runoff from land, especially in the estuarine inlets and deep bays, or the increase of B_{12} may depend on the periodic composition of previous crops such as blooms and seaweed beds." (Provasoli, 1958.)

Algae, as well as planktonic dinoflagellates and diatoms, liberate substances with growth-inhibiting properties for bacteria. It is recognized that, in seawater, bacteria form an important part of the food of invertebrates $(1-10\%$ of the nitrogen requirement of the sediment invertebrates, according to Zhukova (1963)). On the other hand, the marine zooplankton accumulates in zones where the phytoplankton "in full bloom" is fading away. In these zones, the bacteria are increasing as a result of the lack of normal liberation of antibiotic molecules by the phytoplankton. The zooplankton, relying on the bacteria for nutrient and for vitamins then accumulate at the border of the "blooming" phytoplankton.

Amino acids released by zooplankton are also an important source of dissolved organic matter in the sea. They represent an important metabolic pathway in the planktonic ecosystem (Johannes and Webb, 1965). The nutrition of starfish is probably a dual process involving both a continual epidermal absorption of dissolved exogenous materials (amino acids and glucose) for the benefit primarily of the superficial tissues and intermittent oral feeding to satisfy the more general needs of the entire organism and especially of the internal organs (Ferguson, 1967). In soil, the presence of carbohydrate substrates may influence the nutritional behavior of some nematode-trapping fungi. From soil saprophytes they switch to a predaceous activity, only during a relatively short phase when hexoses are available as energy source (Satchuthananthavale and Cooke, 1967). Motile *Escherichia coli* migrate in bands that are influenced by oxygen and organic nutrients such as galactose, glucose, aspartate, threonine, and serine (Adler, 1966).

A biotic community maintains itself by the process of aggregation, through reproduction in the absence of dispersive processes. Marine larvae appear to recognize special substrates and to delay metamorphosis until the suitable conditions are encountered. Barnacle cyprids, for instance, respond to contact with their own species by settling nearby (Knight-Jones and Stevenson, 1950). The settling factor has been shown to be arthropodin, recognized by the larva through a "tactile chemical sense" (Crisp and Meadows, 1963).

In symbiotic associations between autotrophic green plants and heterotrophic organisms, molecules such as carbohydrates have been shown to move from the green plant to the other partner (Richardson *et al.*, 1967). Many ecological aspects should be classified as *molecular reactions* (liberation of molecular factors by organisms in the habitat) and *molecular actions* (infringement of molecular factors upon organisms). In an ecosystem, besides the contribution of the trophic chains in supplying molecules endowed with nutritive or regulatory functions, one may describe molecules active in the constitution and the maintenance of the biotic community. It is convenient to record them under the general denomination of *ecomones*. An ecomone may act in different respects. For instance, the concentration of dissolved carbohydrates in seawater greatly varies from place to place and may have an influence on the nature and growth of the phytoplankton. The ecological importance of these carbohydrates has also been detected with respect to the pumping rate of oysters. Collier (1953) has shown that each oyster has a threshold limit to carbohydrate, above which it will pump, and this threshold is raised with increasing temperature.

Other ecomones are recognized as being specifically active in the process of the coaction of the organisms upon each other. Such specific substances or *coactones* are determinant in the relationship of the *coactor* (active or directing organism) to the *coactee* (passive or receiving organism). This implies the existence of chemosensory functions that are however far from being fully understood. Behavioral studies indicate that the antennae in arthropods, the mantle cavity in mollusks (Kohn, 1961), the skin of Batracians, etc., are the sites of the chemoreception. The electrophysiological data providing the clearest demonstration of the role of these organs are still lacking.

A number of coactones are liberated by the coactor in the medium and reach the coactee. These we may call *exocoactones*. Among these are, for instance, the molecular factors of the orientation of animals through the perception of the odor of the animals taken as good. In the case of the minnow *Hyborhynchus* the ability has been demonstrated to discriminate between three species of invertebrates (*Gammarus fasciatus, Hesperophylax* sp., *Hyalella*) (Hasler, 1954). *Hyborhynchus notatus* has even been trained to distinguish aquatic plants (Walker and Hasler, 1949). The feeding movements of barnacles are known to be correlated with metabolic products of dinoflagellates (Allison and Cole, 1935). Fishes liberate an attractant for sharks and, if frightened, secrete more of it. Therefore only the distressed fish are attacked (Tester, 1963). This interpretation has however been questioned recently. Experiments performed by Kalmijn (1966) demonstrate that sharks detect motion in other animals by sensitive electroperception of their spread muscle potentials. The receptor organ, having the same histological origin as the lateral line system, is the ampulla of Lorenzini. It is filled with a jelly-like material formed by a mucopolysaccharide-protein complex, the carbohydrate moiety of which varies from species to species (Doyle, 1968). This, rather than a chemoreceptor is a better explanation of the extraordinary speed with which sharks can appear on the scene when a fish is seriously disturbed or wounded (Eibl-Eibesfeldt and Hass, 1959).

Specific feeding relationships of insects to plants are identified in many cases at the level of the attractant liberated by the plant, as in the case of *Scolytus multistriatus* responding to a compound produced in the bark of *Ulmus americana* (Norris and Baker, 1967). Essential oils are the main attractants of this category. *Pieris* larvae are attracted by mustard oils. The insects feeding on Rosaceae are attracted by amygdalin. Ammonia and trimethylamine of the cotton plants attract cotton-feeding insects. On the other hand, in the case of the southern armyworm *Prodenia eridania*, known for its polyphagous habits, no specific feeding attractants have been found (Soo Hoo and Fraenkel, 1966). Odorous constituents of decomposing flesh, feces, and urine attract coprophagous and necrophagous species. Many other cases of specific attractants are described by Dethier (1947). Chemical attractants ensure the final and specific guidance toward the food, though the attractant does not necessarily represent the food itself. For instance, the beetle *Creophilus* is attracted toward decaying meat by its odor, but it feeds on fly maggots developing in the meat. Bees are attracted toward flowers by essential oils, but they feed on nectar.

Selection of oviposition sites is often operated through chemical attractants, and the oviposition often happens on the food upon which the insect feeds as a larva. As shown by Traynier (1965), the oviposition of *Erioischia brassicae* is correlated with the presence of mustard-oil glucosides in host plants.

Repellent chemical compounds, of a volatile nature, protect certain plants from attack by insects: this, for example is the case with the teak and cypress pines which are immune to termite attack due to the presence of a sesquiterpene (Oshima, 1919) and for certain pines which are immune to the attack of the nun moth, *Liparis monacha,* due to the presence of turpentine (Sweetman, 1936). 6-Methoxybenzoazolinone is present in corn plants resistant to the corn borer *Pyrausta nubilalis* (Beck, 1960).

In Myriapods, IICN is often used as a repellent (Blum and Woodring, 1962), and mechanisms are at work to keep a stable reserve of cyanogenic compounds (Eisner et al., 1963). In *Archiulus sabulosus* a poison of quinonic nature has been isolated (Trave et al., 1959).

In many insects, chemical repellents are used as protection against enemies (Eisner and Meinwald, 1966)(Fig. 3-1). In this quality, larvae of *Melosoma populi* produce salicylaldehyde (Garb, 1915) and *Carabus* butyric acid (Marchal, 1913). The bombardier beetle *Brachinus* expels a corrosive vapor which is a mixture of p-benzoquinone and toluchiquinone. The secretion of the pygidial glands of this insect contains hydroquinone, hydrotoluquinone, and hydrogen peroxide (Schildknecht and Holoubek, 1961; Schildknecht, 1963).

The cockroach *Eurycotis floridana* repels ants through a liberation of 2-hexenal (Roth et al., 1956) and in *Eurycotis decipiens* a large defensive gland, which opens through the intersegmental membrane between the sixth and seventh sternites, produces D-gluconic acid (Datco and Roth, 1967). The tenebrionids' repellents are quinones (Schildknecht, 1963) and those of *Forficula auricularia* are toluquinone, ethylquinone, and hydroquinone (Schildknecht and Krämer, 1962). Diplopoda liberate toluquinone and methoxytoluquinone (Schildknecht, 1963). The repellents liberated by Pentatomidae (stink bugs) are Δ^2-n-hexenal, Δ^2-n-octenal and Δ^2-n-decenal (Schildknecht and Weis, 1960). *Carabus auratus* uses methacryl acid and tiglic acid as repellents

p-Benzoquinone   Toluquinone   Methyoxytoluquinone

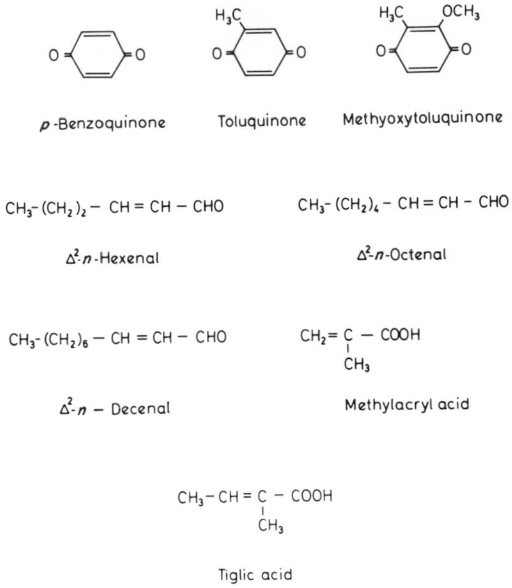

$CH_3-(CH_2)_2- CH = CH - CHO$ $CH_3-(CH_2)_4- CH = CH - CHO$

Δ^2-n-Hexenal Δ^2-n-Octenal

$CH_3-(CH_2)_6- CH = CH - CHO$ $CH_2= \underset{CH_3}{C} - COOH$

Δ^2-n − Decenal   Methylacryl acid

$CH_3-CH = \underset{CH_3}{C} - COOH$

Tiglic acid

FIG. 3-1.   Examples of repellents.

(Schildknecht and Weiss, 1960) as do other species of *Carabus.* When the eggs of the species used as hosts are parasitized by the wasp *Trichogramma evanescens*, they communicate chemical cues to other females and by this reproductive deterrent prevent repeated ovipositions (Salt, 1936).

In *Papilio machaon*, the eversible cervical gland (osmeterium) of the caterpillar produces isobutyric and 2-methylbutyric acids that are effective defense against ants (Eisner and Meinwald, 1965). The mandibular glands of worker honey bees secrete 10-hydroxy-2-decenoic acid, the chief lipoid component of larval food. Its main function is to repel robbers which explains, in the social organization of the hive, its repellency to nectar-collecting bees (Simpson, 1966). Repellents are also present in other animal groups. In *Aplysia*, the repellent aplysioviolin is a monomethyl ester of a biladiene dicarboxylic acid (Rüdiger, 1967), while glomerine, the repellent of *Glomeris marginata* is an alkaloid (Schildknecht *et al.* 1966; Schildknecht and Wenneis, 1966). The main component of the poison produced by the skin of the salamander is also an alkaloid (Habermehl, 1966). Over 1000 species of marine organisms are known to be venomous or poisonous (Russell, 1965). In the case of the helmet conch (*Cassis tuberosa*) a neurotoxin secreted in the saliva is active on numerous marine organisms. It explains why this mollusk can attack and eat the black sea urchin (*Diadema antillarum*) by paralyzing the movements of the prickles (Cornman, 1963). Tetrodotoxine, another neurotropic compound produced by the bowlfish, has been extensively studied (Tsuda, 1966; Kao, 1966).

Parasites are in many cases attracted toward their host by the action of exocoactones. For instance this is the case in nematode parasitizing the roots of plants. It has been shown that glutamic acid is one of the coactones active in this attraction (Bird, 1959, 1960). Young *Petromyzon marinus* are attracted toward the fish they parasitize by a coactone which has been shown to be an amine (Kleerekoper and Mogensen, 1963).

Molecules can be liberated by plants and in one way or another influence the growth of other plants. An example is found in two species of walnut *Juglans regia* and *J. nigra* in which a toxic compound present in the leaves, juglone, is washed out of the leaves by rain and restricts the plants under the leaf canopy (Davis, 1928).

Juglone

The migration of the exocoactone can also take place, not through the air, but through the soil. Here the nature of the biochemical continuum is controlled by the relative proportions of the liberation of organic compounds and their destruction by microorganisms. The concept that these molecular units are too rapidly destroyed to have any action is abandoned, and it is now accepted that coactones from plant or animal origins can act as favorable or unfavorable factors on plants, and even pass from

one plant to another with or without modification during its trans-
location in the soil. These movements of metabolic products
may influence in many ways plants (literature in Grümmer, 1961
and in Winter, 1961) as well as animals. *Protodrilus symbioticus*,
an archiannelid, prefers sand from its habitat in the natural state
to the same sand altered by various physical or chemical treat-
ments (Gray, 1966).

Secondary plant substances may act on insects without being
liberated in the medium: the plant growth substance gibberellic
acid and the insect growth substance ecdysone-λ have similar
effects on both plants and locusts (Carlisle *et al.*, 1963). The feeding behavior when the animal is on the plant on to which it may
have been attracted by other factors may also be unlocked specifically by plant products. Such *endocoactones* are largely responsible for the food specificity of insects. The young larvae of
Bombyx mori are attracted to the leaves of the mulberry tree or of
other plants by an exocoactone, α,β-hexenal (and the older larvae
by β,γ-hexenal) (Watanabe, 1958), but the silkworm is induced to
consume the leaves only in the presence of two other factors
found in mulberry leaves: a biting factor (β-sistosterol and iso-
quercitin or morin) and a swallowing factor (cellulose and the co-
factors sucrose, inositol, inorganic phosphate, and silica)
(Hamamura *et al.*, 1962, 1966). The plant-insect relationship may
even be more complex, as shown by the inducement of the mating behavior in *Polyphemus* moths. Emanation from oak leaves
acts specifically on receptors located on the antennae of the female. This is an absolute prerequisite for the female to release her
sex pheromone that activates the males (Riddiford and Williams,
1967).

In the ecological relationship of plants and insects, it is important to consider, as suggested by Fraenkel (1959), that the primary components of plants, resulting from the biosynthetic pathways also existing in bacteria and animals, are the same in every
plant species. The basic requirements of phytophagous insects
are covered by these constituents. But the plants contain a vast
array of "secondary substances" and Fraenkel suggests that
these (glucosides, saponins, tannins, alkaloids, essential oils,
etc.) have been developed as repellents, unpalatable compounds, or poisons for insects. Had these aspects of adaptive

evolution been fully successful, insects would no longer be dangerous for plants. But according to Fraenkel, insects have, on their part, responded to this chemical control by the plant. When a given insect species, by genetic selection, overcame the repellent effect it gained a new source of food, further selection producing new species attracted by the former repellent. This points to the importance of the reciprocal selective responses in the originization of organic diversity (Ehrlich and Raven, 1964). A remarkable catalog listing secondary plant substances (except alkaloids) by chemical structure and source has been compiled by Karrer (1958). An interesting parallel may be drawn from the facts that many of these compounds are also the active components in the defensive secretion of arthropods, for instance, 2-hexenal produced in the defensive spray of several insects (Roth and Eisner, 1962; Jacobson, 1966) and α-pinene ejected by termites (Moore, 1964) are widespread in plants.

A plant may be attractive and poisonous, but if the insect develops an adaptation to particular secondary components of a particular species of plant, these components may increase the food specialization by their function as endocoactones, such as biting factor, chewing factor, etc. — all these will lead to a consumption of the plant food sufficient to ensure the welfare of the insect.

According to these views, secondary substances of plants are repellents, attractants, or we may add, endocoactones, either making the plant impalatable or even poisonous, or making the plant bitable, chewable, palatable, and therefore consumed in amounts covering the quantitative needs of the insects. This applies not only to the relationship between plants and insects, but also between plants and snails for instance (Stahl, 1888; contested by Frömming, 1962). As shown by Frings and Frings (1965) *Ulva lactuca* produces a compound that acts as an attractant for *Aplysia juliana*. Other effects of endocoactones are beginning to be identified. Glutathione, for instance, acts as a phagostimulant for *Hydra* (Loomis, 1959).

It applies also to the relationship existing between an animal and some of its parasites. The peculiar stimulation of mosquitoes which ensures gorging and subsequent dispatch of the food to the midgut is chemical in nature and comes from the blood cells

(Hosoi, 1959). It is mainly adenosinephosphates that act as stim-
ulant, though the presence of sodium ions is also required
(Galun *et al.*, 1963).

$$CH_3 - \overset{\overset{\displaystyle O}{\|}}{C} - (CH_2)_5 - CH = CH - \overset{\overset{\displaystyle O}{\|}}{C} - OH$$

9 − Ketodec − *trans* − 2 − enoic acid

   According to this conception, food specialization is the result
of the existence of coactones ensuring a sufficient consumption
of a given food and not of special nutritive characteristics of the
food − all living creatures are equivalent in this respect, as a consequence of their remarkable unity of composition with regard to
their primary constituents. This can certainly not be generalized
as representing the nutritional interrelation in all living species.
Some primary constituents of certain organisms may be required
by some other organisms to cover their specialized chemical
needs. It was believed for some time that whales converted the
β-carotene of the krill into vitamin A in the intestine in an amount
sufficient to cover their chemical need in this vitamin (Wagner,
1939). But it was later found that the carotenoid characteristic of
the crustacea, the euphausiids, used by whales as their main
food, was astaxanthin and not β-carotene. Kon and Thompson
(1949) examining the content of the digestive tract of whales captured in the antarctic found large numbers of *Euphausia superba*
with mere traces of β-carotene. The krill, on the other hand, was
rich in vitamin A. As Wagner had studied arctic whales, the observation was renewed on an arctic fin whale, the krill of which,
composed of another Euphausiid, *Meganyctiphanes norvegica*,
contained traces of β-carotene and larger amounts of astaxanthin,
but was rich in vitamin A (Batham *et al.*, 1951). Among Crustacea, euphausiids are exceptional with respect to their vitamin A
content. As they represent the food of whales, the chemical need
of vitamin A is covered in these animals by the substance itself,
while in other mammals it is mainly covered by the use of provi-
tamin A in the form of carotenes.
   In the flagellates that inhabit the gut of *Cryptocercus punctu-*

latus, the wood-eating roach, Cleveland has shown that sexuality depends on the action of the host's molting hormone (references in Cleveland, 1956). A number of correlations between the inducement of sexuality in parasitic Protozoa and the sexual development of the host have been reported. The development of the gregarine *Lankesteria culicis* in mosquitoes shows this relationship (Wenyon, 1911; Ganapati and Tate, 1949). In *Opalina ranarum,* parasitic in rectum of *Rana temporaria,* the induction of the sexual reproductive pattern is associated with the level of the gonadal hormones in the frog (El Mofty and Smyth, 1960). In insects, a synchronization host-parasite has also been found, and the hormonal level of the host that is important in determining the life cycle of the parasite. This has been demonstrated in the case of *Eucarcelia rutilla* (Tachinidae) and its host *Bupalus piniarius* (Geometridae) (Schoonhoven, 1962). Another interesting relation is found when considering the potatoroot eelworm (*Heterodera rostrochiensis* Wollenweber). This species is stimulated to hatch from its cysts by diffusates from the roots of potato, tomato, and other solanaceous plants (Clarke and Widdowson, 1966).

A special category of exocoactones is represented by those substances produced by a coactor and active on a coactee, both coactor and coactee belonging to the same species. These exocoactones are called pheromones. According to Karlson and Lüscher (1959), "Pheromones are defined as substances which are secreted to the outside of an individual and received by a second individual *of the same species,* in which they release a specific reaction, for example a definite behavior or a developmental process."

Sexual hormones, or intraspecific regulators of sexual processes, have been recognized in many fungi (Backus, 1939; Bishop, 1940; Krafczyk, 1935; Raper, 1939, 1940, 1951, 1952, 1954). In phycomycetes, for example, Raper has been able to demonstrate in two heterothallic species of *Achylia* that the sexual process consists of a chain of actions of specific chemical agents, made up of a minimum of seven chemical agents, four secreted by the males and three by the females, as shown in Fig. 3-2.

In some unicellular algae, the immobile eggs secrete soluble substances producing a concentration gradient in the surround-

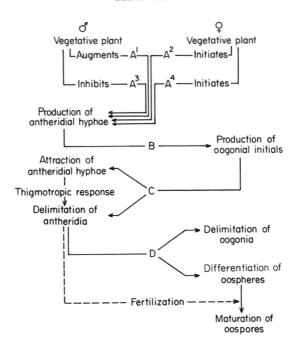

FIG. 3-2. Hormonal mechanism coordinating the sexual process in *Achylia* (Raper, 1951).

ing medium and the antherozoids approach the eggs by swim-ming up the gradient. Pascher (1931 – 1932) has shown this to be the case with *Sphaeroplea* eggs. In certain fucoids, the chemo-tactic agent is volatile (Cook and Elvidge, 1951). Sirenin is the sperm attractant produced by the female gametes of the water mold *Allomyces*; it is an oxygenated sesquiterpene (Machlis *et al.*, 1966). Conjugating and other chemotactic chemicals have been described in many of the fungi (Bonner, 1949; Plempel, 1957, 1963).

As suggested by Karlson (1960), it is useful to distinguish in animals between the pheromones acting olfactorily and those acting orally. Sex attractants, marking scents, and alarm sub-stances belong to the first category.

Alarm or alerting substances are induced by threatening sti-muli and communicate the presence of danger to members of the same species (review in Pfeiffer, 1963; Eisner and Meinwald, 1966). The response is very often flight, for example in the tadpoles of *Bufo*, or in fish (Eibl-Eibesfeldt, 1949; Schutz, 1956; Pfeiffer, 1966a, 1967). The specificity is rather high since the pheromone produced by *Bufo bufo* or *B. calamita* is uneffective on the anurans not belonging to the Bufonidae (Pfeiffer, 1966b). In the boxfish *Ostracion lentiginosus* the alarming substance produced, pahutoxin, has been identified as the choline chloride ester of 3-acetoxyhexadecenoic acid. It is a highly toxic compound able to kill fish. This is so far the first chemical identification of an alarming substance secreted by a marine organism (Boylan and Scheuer, 1967). In gregarious fishes, such as the minnow *Phoxinus phoxinus* L., a small injury to one individual releases a substance producing the rapid flight of the swarm (von Frisch, 1941) and the alarm pheromone has been identified as a purine or a pterine (Hüttel, 1941). Alarm substances of ants generally belong to the category of terpenes with molecular weights between 100 and 200 (Fig. 3-3). The response to the releaser of alarm is generally an attack. Some alarm pheromones are produced by the mandibular glands: dendrolasin in *Lasius (Dendrolasius) fuliginosus* (Pavan, 1961), citral in *Atta rubropilosa* (Butenandt *et al.*, 1959a) and citral and citronellal in *Acanthomyops claviger* (Chadha *et al.*, 1962). Other alerting pheromones are secreted in other ants by anal glands, consisting primarily of methyl-*n*-amyl ketone: methylheptenone in *Tapinoma* (Cavill and Hinterberger, 1961), 2-heptanone in *Iridomyrmex pruinosus* (Blum, 1962; Blum *et al.*, 1963), and isodihydronepetalactone in *Iridomyrmex nitidus* (Cavill and Clark, 1967). The alarm substances in social insects have been reviewed by Maschwitz (1966).

Ants generally migrate on trails indicated by a marking scent laid down by the workers as part of the colony organization (see for instance: Lindauer, 1963; Blum and Ross, 1965; Cavill and Robertson, 1965). This trail pheromone has been identified in *Iridomyrmex detectus* as 2-methylhept-2-en-6-one (Cavill and Ford, 1953). It has been purified from the fire ant *Solenopsis saevissima* but is still unidentified (Walsch and Law, 1965). Trail substances exist also in termites (Becker and Petrowitz, 1967). In bees, each colony has its own smell and the foragers mark interesting spots with a specific scent. Newcomers go to the spots visited by members of their own colony. The odor emitted by the workers is not genetically inherited but derives from metabolic differences between colonies such as food supply, breathing rhythms, etc. It develops even between queenless halves of colonies (Kalmus and Ribbands, 1952). Many mammals release scent to mark out territories and home ranges. Indol and skatol, present in excrement, are used in this function by a number of carnivorous mammals. Other odorants are secreted by exocrine glands located in different regions of the body, according to the species. The wolf rubs his back on the trees limiting his territory and marks them with a specific scent.

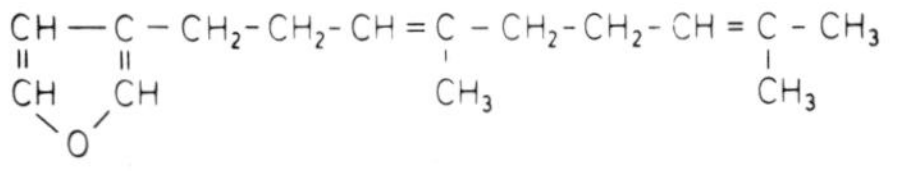

$$CH—C-CH_2-CH_2-CH=C-CH_2-CH_2-CH=C-CH_3$$
$$\underset{CH}{\|}\quad\underset{CH}{\|}\qquad\qquad\overset{|}{CH_3}\qquad\qquad\qquad\overset{|}{CH_3}$$
$$\underset{\searrow O \nearrow}{}$$

Dendrolasin

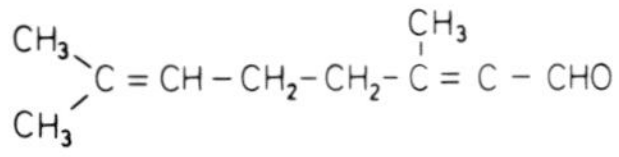

$$\underset{CH_3}{\overset{CH_3}{}}\!\!\!\!C=CH-CH_2-CH_2-\overset{\overset{CH_3}{|}}{C}=C-CHO$$

Citral

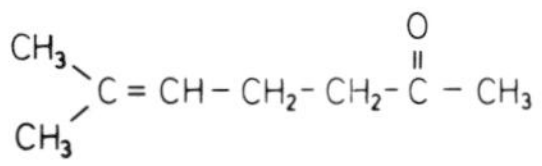

$$\underset{CH_3}{\overset{CH_3}{}}\!\!\!\!C=CH-CH_2-CH_2-\overset{\overset{O}{\|}}{C}-CH_3$$

Methylheptenone

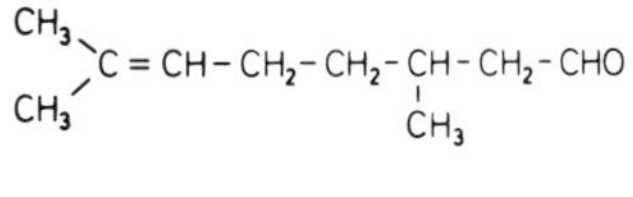

$$\underset{CH_3}{\overset{CH_3}{}}\!\!\!\!C=CH-CH_2-CH_2-\underset{\overset{|}{CH_3}}{CH}-CH_2-CHO$$

Citronellal

$$CH_3-CH_2-CH_2-CH_2-CH_2-\overset{\overset{O}{\|}}{C}-CH_3$$

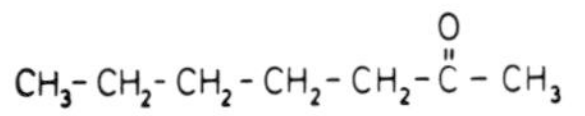

2-Heptanone

FIG. 3-3. Alarm substances of ants.

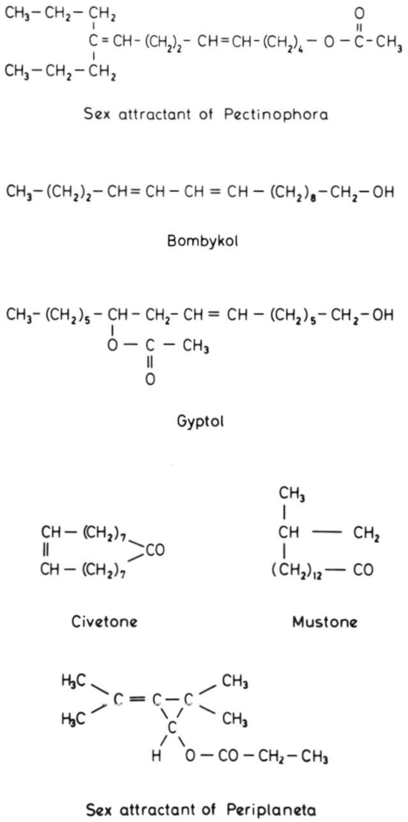

$$CH_3-CH_2-CH_2$$
$$CH_3-CH_2-CH_2$$

$$\overset{CH_3-CH_2-CH_2}{\underset{CH_3-CH_2-CH_2}{C}}=CH-(CH_2)_2-CH=CH-(CH_2)_4-O-\overset{O}{\overset{\|}{C}}-CH_3$$

Sex attractant of Pectinophora

$$CH_3-(CH_2)_2-CH=CH-CH=CH-(CH_2)_8-CH_2-OH$$

Bombykol

$$CH_3-(CH_2)_5-\underset{\underset{O}{\underset{\|}{\overset{|}{C}}-CH_3}}{\overset{|}{CH}}-CH_2-CH=CH-(CH_2)_5-CH_2-OH$$

Gyptol

$$\overset{CH-(CH_2)_7}{\underset{CH-(CH_2)_7}{\|}}\!\!\searrow\!\!CO$$

Civetone

$$\overset{CH_3}{\underset{(CH_2)_{12}-CO}{\overset{|}{\underset{|}{CH}}-CH_2}}$$

Mustone

$$\overset{H_3C}{\underset{H_3C}{}}\!\!\searrow\!\!C=C-\overset{CH_3}{\underset{\underset{H}{\overset{|}{C}}\ \ O-CO-CH_2-CH_3}{C}}\!\!\nearrow\!\!\overset{CH_3}{\underset{CH_3}{}}$$

Sex attractant of Periplaneta

FIG. 3-4.   Sex pheromones.

Sex pheromones (Fig. 3-4), releasing sexual attraction or sexual behavior, or both, are known in many animal taxa (literature in Wilson and Bossert, 1963)—insects (see also Butler, 1967; Jacobson, 1965), crustacea, fish, salamanders, snakes, and mammals, as well as in the lower phyla.

In the rotifer *Brachiorus*, male mating reactions are induced by a substance continuously released in the environment by growing females (Gilbert, 1963). Ants of different sexes and castes produce different odorous compounds (Law *et al.*, 1965). A mixture of terpenes and an indole base is released during the mating flights by the male of *Lasius* sp. and *Acanthomyops* sp. In *Pheidole fallax* Mayr, the soldier ants secrete an indole base, presumably stable, while the minor workers produce a trail substance so far unidentified. In *Bombus terrestris* the male obtains farnesol from essential oils of the flowers and accumulates it in the mandibular glands. It sprays it around to gather the females (Stein, 1963).

Bombykol, the sex attractant of *Bombyx mori*, is an aliphatic molecule with two double bonds and the empirical formula $C_6H_{30}O$. Its structure has been elucidated by Butenandt *et al.* (1959b, 1961a,b; 1963) (Fig. 3-4). A few molecules act on a receptor of the antennae of the male and cause the excitation which unlocks the male behavior.

Gyptol is the sex attractant of the gipsy moth, *Porthetria dispar* (Jacobson *et al.*, 1960). The attractant released by the female of *Periplaneta americana* and acting on the male as excitant and releaser of copulation behavior has been isolated (Wharton *et al.*, 1962) and identified as 2,2-dimethyl-3-isopropylidene-cyclopropyl propionate (Jacobson *et al.*, 1963).

In the frass produced by the male of *Ips confusus* boring in pondera pine (*Pinus ponderosa*), there is a powerful assembling scent evoking the mass attack by individuals of both sexes. The females are however more responsive than the males. When the insects arrive at the source of attraction, they take part in boring, feeding, mating, and oviposition (Wood *et al.*, 1967). Three terpene alcohols have been identified and synthesized (Silverstein *et al.*, 1966). They are inactive individually, but produce a response if two of them are used. The sex attractant in the black carpet beetle, *Attagenus megatoma*, has been identified as the *trans*-3-*cis*-5-tetradecadienoic acid (Silverstein *et al.* 1967). The virgin female in the pink bollworm (*Pectinophora gossypiella*), a cotton pest, produces an attractant whose structure has been identified as 10-propyl-*trans*-5,9-tridecadienyl acetate. It is the first naturally occurring compound with a propyl branching to be reported (Jones *et al.*, 1966).

In some species it has been shown that the male attractant produced by female virgins is under endocrine control of the *corpora allata*. This is the case with some cockroaches as demonstrated by Barth (1965). An interesting observation is that in the parthenogenic strain, the *corpora allata* seem to have lost their ability to control the production of pheromone.

Civetone, the secretion of the para-anal glandular pouch of the civet *Viverra zibetha*, is possibly a sex pheromone, although it is also used in defense. Muskone found in the preputial glands of the musk deer *Moschus moschiferus* probably functions either as a sex attractant, a territorial marker, or both. The prostaglandins present in the semen of mammals and synthesized from unsaturated fatty acids may perhaps act as pheromones in controlling some aspects of the physiology of the uterus (Bergström, 1967). Sexual pheromones of mammals may, as indicated by Lederer (1950), show a relationship with hormonal steroids. This points to a possible function as sex pheromones for the odorant steroids of mammalian urine. The characteristic "urine odor" of humans is partially due to Δ^2-androstene-one-17. Carr and Caul (1962) have been able to train female rats to distinguish the odors from normal versus castrated male rats, and Carr and Pender (1962) have shown that male rats can distinguish the odors of urine from estrous versus diestrous females. Odoriferous substances have been shown to be involved in the social relations of *Petaurus breviceps papuanus* Thomas (Schultze-Westrum, 1964).

There are also pheromones which do not release any complex behavior but act simply as promoting aggregation. For instance, in a species forming clusters, the beetle *Lycus loripes*, a volatile attractant is produced by the males (Eisner and Kafatos, 1962). Pheromones may also act as increasing the species dispersal. Adults of *Tribolium castaneum*, at high density, distribute themselves uniformly as a result of the liberation of 2-ethyl-1,4-benzoquinone and of 2-methyl-1,4-benzoquinone (Loconti and Roth, 1953).

Other pheromones act orally rather than olfactorily. The queen substance of the honey bee is in this category. The chemical structure has been recognized as that of an unsaturated keto acid (Butler *et al.*, 1959; Callow and Johnston, 1960; Barbier *et al.*, 1960).

The information carried to the colony by this substance is the presence of the queen. Ingestion of this substance by the worker bee licking the body of the queen inhibits the development of ovaries in workers and influences their behavior by preventing queen cell construction (see also Rembold, 1964). The queen substance has been reported to be without effect on ovary development in the housefly (Mitlin and Baroody, 1958). However Nayar (1963) was able to contradict this finding by using an appropriate concentration.

In termites, the workers are not specialized adults as is the case for bees, but are larvae capable of metamorphosing into soldiers or reproductives which can substitute for the king and queen. Removal of the king and queen produce the appearance of reproductives by a special molt which, in the presence of the ruling couple, is inhibited by two pheromones produced by them and excreted with their faeces. These being taken up by the workers, again excreted, and again eaten, the pheromones are spread through the whole colony and maintain the social structure. In *Vespa orientalis* F. the social organization is also due to the presence of pheromones (Ishay *et al.*, 1965).

The circulation of molecules and macromolecules through the *biochemical continuum* and the factors controlling their distribution defines life better than any other aspect of it. From what we have said above, it is clear that the nature and economy of any ecosystem depends on a series of messages controlling the metabolism of the biotic community in a given environment. A number of these messages originate from the very physical nature of the environment, but it is now clear that inside the network of the *biochemical continuum*, there is a circulation of specific molecules or macromolecules endowed with a certain amount of information. The chemical properties of a number of these chemical messengers, essential for the maintenance of the biotic community considered, are already known. In many, though not in all cases, the molecules of coactones are products of lateral extensions of the biosynthetic mevalonic pathway, and it would be difficult to deny their adaptive nature. This implies two different aspects: the production and liberation of the specific coactone by the coactor and the reception of the message by the coactee. The release of the coactone, its rate of spread, and its persistance have been studied quantitatively by considering four general cases (Bossert and Wilson, 1963): (a) instantaneous release of the substance in still air, (b) continuous release in still air, (c) continuous release from a moving source as in odor trail, and (d) continuous release in a wind.

The mathematical approach, based on a diffusion-kinetic model, described successfully the range of the signal as well as other important characteristics of olfactory communication systems. The hypothesis has been advanced that low-frequency molecular vibrations or rotations provide the physical basis of odor and Wright (1963) has elaborated a theory based on the mechanical properties of the molecules which are demonstrably perceptible to various insect species. As to the nature of the reception mechanism, all has still to be done. The extremely small amount of substance necessary to produce a response from the coactee renders the task very arduous. In the case of sex attractants produced by the female of arthropods, 10^{-17} gm of the compound, or even less, are still effective on the male (Wharton et al., 1962).

This problem is essentially the same as that dealt with in permeability studies when considering the chemical nature of the active sites in the membranes (Schoffeniels, 1967). It remains largely unexplained although representing one of the main factors in ecophysiological relations at the molecular level. The most remarkable feature of the studies concerned with the chemical regulation of courtship and other social activities in invertebrates and more specifically in insects is that they outdate the philosophical arguments of psychological thesis luring the lazy mind with a dangerous word: the instinct. They definitively show that a highly sophisticated social organization such as that exhibited by a colony of bees finds its explanation at the molecular scale of dimensions.

References

Adler, J. (1966). *Science* **153**, 708.
Allison, J. B., and Cole, W. H. (1935). *Bull. Biol. Lab. Mt. Desert Island*, **24**.
Backus, M. P. (1939). *Bull. Torrey Botan. Club* **66**, 63.
Barbier, M., Lederer, E., Reichstein, T., and Schindler, O. (1960). *Helv. Chim. Acta*, **43**, 1682.

Barth, R. H. (1965). *Science,* **149,** 882.

Batham, E., Fisher, L. R., Henry, K. M., Kon, S. K., and Thompson, S. Y. (1951). *Biochem. J.* **48,** x.

Beck, S. D. (1960). *Ann. Entomol. Soc. Am.* **53,** 206.

Becker, G., and Petrowitz, H. J. (1967). *Naturwissenschaften* **54,** 16.

Bergström, S. (1967). *Science.* **157,** 382.

Bird, A. F. (1959). *Nematologica* **4,** 322.

Bird, A. F. (1960). *Nematologica* **5,** 217.

Bishop, H. (1940). *Mycologia* **32,** 505.

Blum, M. S. (1962). (personal communication to Wilson and Bossert, 1963).

Blum, M. S., and Ross, G. N. (1965). *J. Insect Physiol.* **11,** 857.

Blum, M. S., and Woodring, J. P. (1962). *Science* **138,** 512.

Blum, M. S., Warter, S. L., Monroe, R. S., and Chidester, J. C. (1963). *J. Insect Physiol.* **9,** 881.

Bonner, J. T. (1949). *J. Exptl. Zool.* **110,** 259.

Bossert, W. H., and Wilson, E. O. (1963). *J. Theoret. Biol.* **5,** 443.

Boylan, D. B., and Scheuer, P. J. (1967). *Science* **155,** 52.

Butenandt, A., Linzen, B., and Lindauer, M. (1959a). *Arch. Anat. Microscop. Morphol. Exptl.* **48,** 13.

Butenandt, A., Beckmann, R., Stamm, D., and Hecker, E. (1959b). *Z. Naturforsch.* **14b,** 283.

Butenandt, A., Beckmann, R., and Hecker, E. (1961a). *Z. Physiol. Chem.* **324,** 71.

Butenandt, A., Beckmann, R., and Stamm, D. (1961b). *Z. Physiol. Chem.* **324,** 84.

Butenandt, A., Hecker, E., and Zayed, S. N. A. D. (1963). *Z. Physiol. Chem.* **333,** 114.

Butler, C. G. (1967). *Biol. Rev.* **42,** 42.

Butler, C. G., Callow, R. K., and Johnston, N. C. (1959). *Nature* **184,** 1871.

Callow, R. K., and Johnston, N. C. (1960). *Bee World* **41,** 152.

Calvin, M. (1962). *Perspectives Biol. Med.* **5,** 147.

Carlisle, D. B., Osborne, D. J., Ellis, P. E., and Moorhouse, J. E. (1963). *Nature* **200,** 1230.

Carr, W. J., and Caul, W. F. (1962). *Animal Behaviour* **10,** 20.

Carr, W. J., and Pender, B. (1962). [cited by Wilson and Bossert, *Recent Progr. Hormone Res.* **19,** 673 (1963)].

Cavill, G. W. K., and Clark, D. V. (1967). *J. Insect Physiol.* **13,** 131.

Cavill, G. W. K., and Ford, D. L. (1953). *Chem. Ind.* **35,** 351.

Cavill, G. W. K., and Hinterberger, H. (1961). *Verhandl. 11 Intern. Kongr. Entomol. Wien 1960.* **III,** 53 – 59.

Cavill, G. W. K., and Robertson, P. L. (1965). *Science* **149,** 1337.

Chadha, M. S., Eisner, T., Monro, A., and Meinwald, J. (1962). *J. Insect Physiol.* **8,** 175.

Clarke, A. J., and Widdowson, E. (1966). *Biochem. J.* **98,** 862.

Cleveland, L. R. (1956). *J. Protozool.* **3,** 161.

Collier, A. (1953),. *Trans. 18th Am. Wildlife Conf.* p. 463

Cook, A. H., and Elvidge, J. A. (1951). *Proc. Roy. Soc. London* **B138,** 97.

Cornman, J. (1963). *Nature* **200,** 88.

Crisp, D. J., and Meadows, P. S. (1963). *Proc. Roy. Soc. London* **B158**, 364.
Datco, G. P., and Roth, L. M. (1967). *Science* **155**, 88.
Davis, E. F. (1928). *Am. J. Botany* **15**, 620.
Dethier, V. G. (1947). "Chemical Insect Attractants and Repellents." Blakiston, Philadelphia.
Doyle, J. (1968). *Comp. Biochem. Physiol.* **24**, 479.
Ehrlich, P. R., and Raven, P. H. (1964). *Evolution* **18**, 586.
Eibl-Eibesfeldt, I. (1949). *Experientia*, **5**, 236.
Eibl-Eibesfeldt, I. and Hass, H. (1959). *Z. Tierpsych.* **16**, 733.
Eisner, T., and Kafatos, F. C. (1962). *Psyche* **69**. 53.
Eisner, T., and Meinwald, J. C. (1965). *Science* **150**, 1733.
Eisner, T., and Meinwald, J. C. (1966). *Science* **153**, 1341.
Eisner, T., Eisner, H. E., Hurst, J. J., Kafatos, F. C., and Meinwald, J. (1963). *Science*, **139**, 1218.
El Mofty, M., and Smyth, J. D. (1960). *Nature* **186**, 559.
Ferguson, J. C. (1967). *Biol. Bull.* **132**, 161.
Fox, S. W., and Yuyama, S. (1963). *Ann. N.Y. Acad. Sci.* **108**, 487.
Fox, S. W., McCauley, R. J., and Wood, A. (1967). *Comp. Biochem. Physiol.* **20**, 773.
Fraenkel, G. (1959). In "Biochemistry of Insects" (L. Levenbook, ed.) pp. 1—14. Pergamon, London.
Frings, H., and Frings, C. (1965). *Biol. Bull.* **128**, 211.
Frömming, E. (1962). "Das Verhalten unserer Schnecken zu den Pflanzen ihrer Umgebung." Duncken and Humblot, Berlin.
Galun, K., Avi-dor, Y., and Bar-Geev, N. (1963). *Science* **142**, 1674.
Ganapati, P. N., and Tate, P. (1949). *Parasitology* **39**, 291.
Garb, G. (1915). *J. Entomol. Zool.* **7**, 87.
Gilbert, J. J. (1963). *J. Exptl. Biol.* **40**, 625.
Granick, S. (1965). In "Evolving Genes and Proteins" (V. Bryson and H. J. Vogel, eds.) p. 67. Academic Press, New York.
Gray, J. S. (1966). *J. Marine Biol. Assoc. U. K.* **46**, 627.
Grümmer, G. (1961). In "Mechanisms in Biological Competition" *Symp. Soc. Exptl. Biol.* No. 15, pp. 219—229. Cambridge Univ. Press.
Habermehl, G. (1966). *Naturwissenschaften* **53**, 123.
Hamamura, Y., Hayashiya, K., Naito, K., Matsuura, K., and Nishida, J. (1962). *Nature* **194**, 754.
Hamamura, Y., Kuwata, K., and Masuda, H. (1966). *Nature* **212**, 1386.
Harvey, H. W. (1937). *J. Marine Biol. Assoc. U. K.* **22**, 97.
Hasler, A. D. (1954). *J. Fisheries Res. Board Can.* **11**, 107.
Hayes, J. M. (1967). *Geochim. Cosmochim. Acta* **31**, 1395.
Hill, C. H., Starcher, B., and Kim, C. (1967). *Federation Proc.* **26**, 129.
Hosoi, T. (1959). *J. Insect Physiol.* **3**, 191.
Hüttel, R. V. (1941). *Naturwissenschaften* **29**, 333.
Ishay, I., Ikan, R., and Bergmann, E. D. (1965). *J. Insect Physiol.* **11**, 1307.
Jacobson, M. (1965). "Insect Sex Attractants." Wiley, New York.

Jacobson, M. (1966). *Ann. Rev. Entomol.* **11**, 403.

Jacobson, M., Beroza, M., and Jones, W. A. (1960). *Science* **132**, 1011.

Jacobson, M., Beroza, M., and Yamamoto, R. T. (1963). *Science* **139**, 48.

Johannes, R. E. and Webb, K. L. (1965). *Science* **150**, 76.

Jones, W. A., Jacobson, M., and Martin, D. E. (1966), *Science* **152**, 1516.

Jukes, T. H. (1966). "Molecules and Evolution." Columbia Univ. Press, New York.

Kalmijn, A. J. (1966). *Nature* **212**, 1232.

Kalmus, H., and Ribbands, C. R. (1952). *Proc. Roy. Soc. London* **B140**, 50.

Kao, C. Y. (1966). *Pharmacol. Revs.* **18**, 997.

Karlson, P. (1960). *Ergeb. Biol.* **22**, 212.

Karlson, P., and Lüscher, M. (1959). *Nature* **183**, 55.

Karrer, W. (1958). "Konstitution und Vorkommen der organischen Planzenstoffe." Birkhauser, Basel.

Kleerekoper, H., and Mogensen, J. (1963). *J. Physiol. Zool.* **36**, 347.

Knight-Jones, E. W., and Stevenson, J. P. (1950). *J. Marine Biol. Assoc. U. K.* **29**, 281.

Kohn, A. J. (1961). *Am. Zoologist* **1**, 291.

Kon, S. K., and Thompson, S. Y. (1949). *Arch. Biol.* **24**, 233.

Krafczyk, H. (1935). *Beitr. Biol. Pflanz.* **23**, 349.

Law, J. H., Wilson, E. O., and McCloskey, J. A. (1965). *Science* **149**, 544.

Leach, R. M. (1967). *Federation Proc.* **26**, 118.

Lederer, E. (1950). *Fortschr. Chem. Org. Naturstoffe* **6**, 87.

Lindauer, M. (1963). *Fortschr. Zool.* **16**, 58.

Lipman, F. (1965). *In* "The origins of prebiological Systems and their molecular Matrices" (S. W. Fox, ed.) p. 259. Academic Press, New York.

Loconti, J. D., and Roth, L. M. (1953). *Ann. Entomol. Soc. Am.* **46**, 281.

Loomis, W. F. (1959). *Ann. N. Y. Acad. Sci.* **77**, 73.

Machlis, L., Nutting, W. H., Williams, W., and Rapoport, H. (1966). *Biochemistry* **5**, 2147.

Marchal, P. (1913). *In* "Dictionnaire de Physiologie" Vol. IX, pp. 73–86. Alcan, Paris.

Maschwitz, U. W. (1966). *Vitamins Hormones* **24**, 267.

Miller, S. L. (1953). *Science* **117**, 528.

Miller, S. L. (1957). *Biochim. Biophys. Acta* **23**, 480.

Mitlin, N., and Baroody, A. M. (1958). *J. Econ. Entomol.* **51**, 384.

Moore, B. P. (1964). *J. Insect Physiol.* **10**, 371.

Nayar, J. K. (1963). *Nature* **197**, 923.

Norris, D. M., and Baker, J. E. (1967). *J. Insect Physiol.* **13**, 955.

Oro, J. (1963). *Ann. N. Y. Acad. Sci.* **108**, 464.

Oshima, M. (1919). *Philippine J. Sci.* **15**, 319.

Pascher, A. (1931–1932). *Jahrb. Wiss. Botan.* **75**, 551.

Pavan, M. (1961). *Atti Accad. Nazl. Ital. Entomol. Rend.* **8**, 228.

Pfeiffer, W. (1963). *Experientia* **19**, 113.

Pfeiffer, W. (1966a). *Naturwissenschften* **53**, 565.

Pfeiffer, W. (1966b). *Z. Vergleich. Physiol.* **52**, 79.

Pfeiffer, W. (1967). *Naturwissenschaften* **54**, 177.

Plempel, M. (1957). *Arch. Microbiol.* **26**, 151.

Plempel, M. (1963). *Naturwissenschaften* **50**, 1226.

Plunkett, M. A., and Rakestraw, N. W. (1955). *Papers Marine Biol. Oceanog. Deep Sea Res.* **3** Suppl., 12.

Ponnamperuma, C. (1965). *In* "The origins of prebiological systems" (S. W. Fox, ed.) p. 221. Academic Press, New York.

Provasoli, L. (1958). *In* "Perspectives in Marine Biology" (A. A. Buzzati-Traverso, ed.). Univ. Calif. Press, Berkeley.

Pullman, B., and Pullman, A. (1962). *Nature* **196**, 1134.

Pullman, B., and Pullman, A. (1963). "Quantum Biochemistry." Wiley (Interscience), New York.

Raper, J. R. (1939). *Am. J. Botany* **26**, 639.

Raper, J. R. (1940). *Am. J. Botany* **27**, 162.

Raper, J. R. (1951). *Am. Scientist* **39**, 110.

Raper, J. R. (1952). *Botan. Rev.* **18**, 447.

Raper, J. R. (1954). *In* "Sex in Microorganisms" (Weinreich, Lewis and Raper, eds.) p. 42. A.A.A.S., Washington.

Rembold, H. (1964). *Naturwissenschaften* **51**, 49.

Richardson, D. H. S., Smith, D. C., and Lewis, D. H. (1967). *Nature* **214**, 879.

Riddiford, L. M. and Williams, C. M. (1967). *Science* **155**, 589.

Roth, L. M. and Eisner, T. (1962). *Ann. Rev. Entomol.* **7**, 107.

Roth, L. M., Neigisch, W. D., and Stahl, W. H. (1956). *Science* **123**, 670.

Rüdiger, W. (1967). *Z. Physiol. Chem.* **348**, 129.

Russell, F. E. (1965). *Advan. Marine Biol.* **3**, 255.

Salt, G. (1936). *J. Exptl. Biol.* **13**, 363.

Satchuthananthavale, V., and Cooke, R. C. (1967). *Nature* **214**, 321.

Schildknecht, H. (1963). *Angew. Chem.* **75**, 762.

Schildknecht, H., and Holoubek, K. (1961). *Angew. Chem.* **73**, 1.

Schildknecht, H., and Krämer, H. (1962). *Z. Naturforsch.* **17b**, 701.

Schildknecht, H., and Weis, K. H. (1960). *Z. Naturforsch.* **15b**, 755.

Schildknecht, H., and Wenneis, W. F. (1966). *Z. Naturforsch.* **21b**, 552.

Schildknecht, H., Wenneis, W. F., Weis, K. H., and Maschwitz, U. (1966). *Z. Naturforsch.* **21b**, 121.

Schoffeniels, E. (1967). "Cellular Aspects of Membrane Permeability." Pergamon, Oxford.

Schoonhoven, L. M. (1962). *Arch. Neerl. Zool.* **15**, 111.

Schultze-Westrum, T. (1964). *Naturwissenschaften* **51**, 226.

Schutz, F. (1956). *Z. Vergleich. Physiol.* **38**, 84.

Silverstein, R. M., Rodin, J. O., and Wood, D. L. (1966). *Science* **154**, 509.

Silverstein, R. M., Rodin, J. O., Burkholder, W. E., and Gorman, J. E. (1967). *Science* **157**, 85.

Simpson, J. (1966). *Nature* **209**, 531.

Singer, M. F. (1967). *Science* **158**, 1550.

Soo Hoo, C. F., and Fraenkel, G. (1966). *J. Insect Physiol.* **12**, 693.

Stahl, E. (1888). *Jena Z. Naturw. Med.* **22**, N. F. XV, 557.

Stein, G. (1963). *Naturwissenschaften* **50**, 305.

Sweetman, H. L. (1936). "The Biological Control of Insects." Comstock, Ithaca, New York.

Tester, A. (1963). *Pacific Sci.* **17**, 145.

Trave, R., Garanti, L., and Pavan, M. (1959). *Chim. Ind.* **41**, 19.

Traynier, R. M. M. (1965). *Nature* **207**, 218.

Tsuda, K. (1966). *Naturwissenschaften* **53**, 171.

Vallentyne, J. R. (1957). *J. Fisheries Res. Board Can.* **14**, 33.

Van Niel, C. B. (1956). *In* "The Microbe's contribution to Biology" (A. J. Kluyver and C. B. Van Niel, eds.). Academic Press, New York.

von Frisch, K. (1941). *Z. Vergleich. Physiol.* **29**, 46.

Wagner, K. H. (1939). "Vitamin A und β-carotin der Finn-, Blau- und Sperm-wals." Barth, Leipzig.

Walker, T. J., and Hasler, A. D. (1949). *Physiol. Zool.* **22**, 45.

Walsch, C. T., and Law, J. H. (1965). *Nature* **207**, 320.

Watanabe, T. (1958). *Nature* **182**, 325.

Watson, J. D., and Crick, F. H. C. (1953). *Nature* **171**, 737.

Wenyon, C. (1911). *Parasitology* **4**, 273.

Wharton, D. R. A., Black, E. D., Merrit, C., Wharton, M. L., Bazinet, M., and Walsch, J. T. (1962). *Science* **137**, 1062.

Wilson, E. O., and Bossert, W. H. (1963). *Recent Progr. Hormone Res.* **19**, 673.

Winter, A. G. (1961). "Mechanisms in Biological Competition." *Soc. Exptl. Biol. Symp.* No. 15, pp. 229–244. Cambridge Univ. Press.

Wood, D. L., Stark, R. W., Silverstein, R. M., and Rodin, J. O. (1967). *Nature* **215**, 206.

Wright, R. H. (1963). *Nature* **198**, 455.

Zhukova, A. L. (1963). *In* "Symposium on Marine Microbiology" (C. H. Oppen-heimer, ed.) pp. 699–710. Thomas, Springfield, Massachusetts.

CHAPTER IV

Chemical Properties of Organisms Related to Physical Properties of the Environment

When considering an organism in its natural environment, the chemical comparative physiologist of the nineteenth century was inclined to emphasize corresponding parameters in the organism and the medium: osmotic pressure in the blood of an aquatic animal and in the medium or temperature inside and outside the organism. If this approach has led to the definition of useful concepts such as homeo- and poecilosmoticity, homeo- and poecilothermia, it has however confused the issue by putting the emphasis on correlation that could not be held responsible for or explain, as was proven later, the inclusion of a certain species in a given environment. In the case of the osmotic relations that an organism establishes with its surroundings, it was generally assumed that one is dealing with the response of the organism to a colligative property of the medium. This is however far from being true since the chemical nature of the constituents of the aquatic medium plays a determinant role in the adaptation of an animal to a given environment.

With respect to thermal energy or light or any other electromagnetic vibration, the matter is different since one deals with a truly physical property of the environment that affects the chemical properties of organisms.

Thermal Energy

It is not so far back in the history of science that temperature was considered to be a measurement of heat. If a rise in temperature is an indication that heat is introduced in a system it is by no means a measure of the amount of energy added unless the amount and kind of substance heated is specified. As we know today, temperature may be regarded as the potential or intensity factor of thermal energy while entropy is the capacity factor.

Increase of temperature almost invariably increases the velocity of a chemical reaction. As shown by Arrhenius (1889), there is a linear relation between the log of the specific rate of the reaction and the reciprocal of the absolute temperature. The slope of the straight line is equal to $-E/R$ where E is the activation energy and R the gas constant. E is the energy the reactants must possess to form the activated complex or transition state. As stated in the theory of absolute reaction rates it is possible to express the equilibrium constant of a reaction in terms of the free energy of activation in the standard state. It follows that an entropy and an enthalpy of activation may also be defined. However the most significant conclusion that can be drawn is that the free energy of activation determines the rate of a reaction at a given temperature. The higher the free energy of activation, the higher the temperature must be for the reaction to proceed at a significant rate. Enzymes, like any other catalysts, decrease very markedly the free energy of activation, which explains why the biochemical processes, characteristics of life, do take place at significant rates at ordinary temperature.

The organisms live generally within a narrow range of temperature variations. Only few organisms can endure temperature above $50-55°C$ or below freezing point. In any case the temperature limits for growth and development of most species are rather narrow as shown by the observations reported in Table 4-1 and it has been estimated that if the average daily temperature prevailing at the surface of the earth were suddenly increased or decreased by only $20°C$, all life would disappear.

Since the energy metabolism of most organisms is geared to catenary reactions involving oxygen as electron acceptor, most of the studies dealing with the effect of temperature on organisms take into consideration the oxygen consumption as truly representing the metabolic state of the organism in a given set of experimental conditions. This however gives a rather crude estimate of a situation that is difficult to interpret at the molecular

TABLE 4-1

REPRESENTATIVE TEMPERATURE LIMITS FOR GROWTH
AND DEVELOPMENT[a]

Organism	Minimum temperature (°C)	Maximum temperature (°C)
Alligator (*Alligator mississipiensis*)	4	39
European lizard (*Lacerta agilis*)	−4	44
Garter snake (*Thamnophis radix*)	0	41
Painted turtle (*Pseudemys elegans*)	1	46
Green frog (*Rana clamitans*)	−0.5	22
Common toad (*Bufo bufo*)	−1	33
Western pine beetle		
(*Dendroctonus varivestis*)	4−7	37.8−40.6
Boll weevil (*Anthonomus grandis*)	−4 to 3	50−60
Housefly (*Musca domestica*)	6.7−7.2	44.6
Bedbug (*Cimex lectularis*)	7.5	34−43.5
Ancylostoma caninum (Nematode)	15	37
Trichuris trichiura (Nematode)	−9 to −12	54
Achromobacter ichthyodermis	−2	30
Corynebacterium diphtheriae	15	40
Clostridium botulinum	18	55
Bacillus subtilis	15	55
Stenothermal thermophils	35	75−85
Myxophyta[b]	−17	85
Rhodophyta[b]	−20	35
Chlorophyta[b]	−20	55
Phaeophyta[b]	−20	35
Arctic fir	−40	30
	(survives at −60°C)	
Date palm	10	54

[a]From Allen, 1950.

[b]No single species has the wide limits shown here. The figures indicate the range possible for the group. (Data from W. S. Spector, 1956.)

scale of dimensions. As shown by the following examples, it cannot enlighten the molecular aspects of the changes that organisms may undergo under the influence of temperature variations.

It has long been known that certain poikilothermal organisms from the polar regions exhibit rates of locomotor activities similar to that of species from tropical regions. This has led many authors to consider that animals from colder regions should have higher metabolic rates than animals from warmer habitats when examined at the same temperature. It has indeed been found that if some poikilothermal animals metabolize at lower rates in colder regions or in colder seasons, other forms known as the "adjusters" possess homeostatic mechanisms compensating for different temperatures. Thus by studying many tropical and arctic groups it has been shown that when placed at the same temperature, the oxygen consumption of cold-water species is greater than that of the warm-water species (Scholander *et al.* 1953). It should however be emphasized that this statement is based on observations made by considering species sometimes far remote on the phyletic tree and without consideration of size. More significant would be results dealing with one group of animals continuously distributed over a wide range of latitude. Unfortunately, this type of information is at the present time very restricted. In the case of polychaete species *Clymenella torquata* and *C. mucosa*, measurements of oxygen consumption indicate genetic divergence in latitudinally separated populations (Mangum, 1963), in agreement with the classical view. The effects of divergence are however masked by a concomitant trend in reduction of size with latitude.

The data collected by Vernberg and Vernberg (1966) on populations of crab of the genus *Uca* indicate that it is impossible to correlate the results solely on the basis of adaptation to habitat temperature. They determined metabolism-temperature (M:T) acclimation patterns for whole animals and tissues of populations of *U. rapax* from Salvador (13°S) and Santos (24°S) in Brazil and of the only temperate zone species in South America, *U. uruguayensis,* from Santos, Florianopolis (28°S), and Torres (29°S) Brazil. These data were compared to those obtained for populations of *U. rapax* from Florida (30°N), Jamaica, and Puerto

Rico (18°N) and for the closely related north temperate zone species, *U. rapax*, from North Carolina (35°N). Above 25°C, *U. pugnax* from the southern hemisphere has a higher metabolic rate than either of the tropical or temperate zone northern hemisphere forms. In the Salvador population, cold-acclimated animals had a higher metabolic rate than warm-acclimated animals at a low temperature, and the warm-acclimated animals had a higher rate than cold-acclimated animals at the higher temperatures. From these data and others obtained on isolated tissues (supraesophageal ganglion and muscle), the authors conclude that the M:T curve in widely separated populations of fiddler crabs does not appear to be simply related to environmental temperature.

Seasonal adaptation in the M:T relationship has been demonstrated in *Emerita talpoida*, an intertidal sand crab. At temperatures below 20°C, the O_2 consumption was greater in winter than in summer. Allowing for differences in size, the winter metabolic rate at 3°C was about four times greater than the summer rate at the same temperature and the winter metabolism at 3°C equals that at 15°C in summer (Edwards and Irving, 1943a). Thus, at low temperatures in winter, the metabolism is maintained at an increased level compared to that in summer.

In sharp contrast, the amphipod, *Talorchestia megalophtalma*, living semiterrestrially on the same beaches as *Emerita*, was reported to show no change of O_2 consumption from winter to summer at the same temperature (Edwards and Irving, 1943b). This was presumably accounted for by the fact that, while growth and activity continue in *Emerita* during the winter, *Talorchestia* remains inactive and apparently hibernates at low temperatures as does *Oniscus asellus* (Edwards, 1946). However, if these same data are replotted to take size differences into account, *Talorchestia* also shows some acclimation between winter and summer (Rao and Bullock, 1954).

In the opinion of Scholander *et al.* (1953), metabolic adjustment is, in temperature-adapted forms, the result of a lateral shift of the M:T curve along the abscissa, and not of a change of slope. But other evidence points to the fact that the latter may also be involved in such adaptations. Thus smaller individuals of a species have been reported to respond to temperature changes more markedly than larger ones (Prosser, 1950). Among other data, this is based on the observations on *Talorchestia* referred to above (Edwards and Irving, 1943b). But on replotting these results logarithmically, rate against temperature, Rao and Bullock (1954) find evidence for an increase of response with size. Calculating the temperature coefficient Q_{10} of the oxygen consumption (i.e., the change in rate observed for each 10° rise in temperature) for the various size groups, they conclude that *Talorchestia* shows increasing Q_{10} with size between 12° and 22°C, although a decrease is observed in winter animals. Again, in *Emerita*, they note an increasing Q_{10} with size above 500 mg and between 16° and 21°C.

The crab *Pachygrapsus crassipes* also shows a definite increase of Q_{10} with size between 16° and 23.5°C but not between 8° and 16°C (Roberts, 1957a b). In the fiddler crab *Uca pugnax*, individuals from New York, North Carolina, and Florida show a decrease in Q_{10} with increasing size, whereas the opposite is observed in forms living at the lower latitude of Trinidad (Tashian, 1956).

The above facts show that some animal species can compensate to a certain extent the effect of temperature changes on their metabolism. This compensation may be merely the result of a simple shift of the M:T curve along the abscissa, but a change in the slope of the curve, i.e., in the temperature coefficient (Q_{10}), may also intervene, as in *Daphnia* (Brown, 1929) and *Uca pugnax* (Tashian, 1956). Obviously other methods of compensation may play an additional part (Bullock, 1955). These may involve more complex changes in the shape of the curve rather than simply in its slope. Whether these homeostatic mechanisms depend on the whole organism, or on certain organs or cells, parts of cells, or enzymes remains to be demonstrated. In *Pachygrapsus* the muscle tissue was found to have been acclimated, while the brain showed no sign of such effect (Roberts, 1957a,b). Respiratory rates of washed liver mitochondria isolated from torpid and aroused bats, as well as from control and cold-acclimated rats, do not indicate any differential effects of cold exposure.

With unwashed mitochondria differences in species occur at higher temperatures; they seem to reflect the influence of heat-labile extramitochondrial factors involved in the bat succinate metabolic pathway (Horwitz and Nelson, 1968).

Temperature stresses can bring about alterations in the biochemical steady states. For instance, the concentration of 15 free amino acids in the intracellular pool of amino acids in muscles of *Eriocheir sinensis* living in fresh water shows a marked drop in free proline concentration in animals kept in $1-3°C$ in comparison with those kept at 15°C. This proline decrease is more pronounced after $6-8$ weeks than after 2 weeks (Duchâteau and Florkin, 1955). Thermal stress also increases hemoglobin synthesis in *Daphnia* if the temperature is raised at constant O_2 pressure. Another effect of temperature can be observed on some aspects of the metabolism underlying osmoregulation. For instance, *Gammarus duebeni* may colonize waters with high salt concentration if their temperature is not more than 4° to 16°C (optimum 6°C) (Kinne, 1956). On the other hand, some marine shrimps may become adapted to brackish water, but only at the high end of their temperature range (Panikkar, 1940).

Acclimation of fish at various temperatures involves modifications of the nervous system which permit its normal functioning in the new thermal conditions. In the goldfish *Carassius auratus* L., this results from an increase in the degree of unsaturation of fatty acids of twenty carbons or more (Johnston and Roots, 1964)—an observation in general agreement with the classical view according to which the degree of unsaturation of the lipid increases at lower temperature (Fraenkel and Hopf, 1940; Hoar and Cottle, 1952; Fawcett and Lyman, 1954).

In the case of a hibernating species such as the Greek tortoise (*Testudo hermanni* Gmelin), seasonal variations have been observed in the permeability characteristics of the intestinal epithelium as well as in the composition of the blood (Table 4-2). It has been shown that the osmotic pressure and the concentration in Na, Cl, and urea increases progressively from the month of August to reach a maximum at the end of the hibernating period, i.e., in April. The aminoacidemia follows the reverse pattern since the highest value is observed during the month of July. If one studies the composition of a tissue such as the intestinal epithelium, it appears, as demonstrated by the results of Gilles-

TABLE 4-2

SEASONAL VARIATIONS IN THE BLOOD COMPOSITION OF THE TORTOISE *Testudo hermanni* GMELIN[a]

Month (Number of animals analyzed)	Na (mEq/l)	K (mEq/l)	Ca (mEq/l)	Cl (mEq/l)	Urea (mM/l)	Osmotic pressure (mOsm/l)		Amino acids nitrogen (mM/l)
						Measured	Calculated	
April 1964 (2)	167.1± 3.7	4.60±0.23	4.58±0.08	134.0±12.0	103.7± 9.8	467.2± 19.0	415.5±24.0	1.2
March 1964 (1)	129.2	4.99	4.90	86.6	37.3	340.6	263.0	2.5
June 1964 (1)	105.2	4.25	1.50	66.8	26.6	258.5	204.4	1.9
July 1964 (3)	115.4± 6.3	4.47±0.34	4.53±1.87	94.7± 5.2	3.9± 0.7	290.2± 5.5	223.1± 7.2	2.9
August 1964 (1)	135.8	4.81	5.50	108.7	12.4	322.9	267.2	—
September 1964 (5)	135.7± 9.4	4.13±1.14	4.90±4.30	99.0±23.0	11.2± 8.2	338.6± 40.1	254.3±25.4	1.6
October 1964 (5)	138.3± 5.2	4.21±0.60	4.81±2.11	110.3±25.5	22.7±14.1	343.7± 20.5	280.3±37.1	1.7
November 1964 (2)	141.1± 3.3	2.98±0.57	5.15±0.25	99.4± 4.0	21.9± 2.0	349.1± 2.1	270.6± 1.0	—
December 1964 (3)	155.9± 1.9	3.87±0.23	6.33±1.67	124.6±13.3	31.9± 2.9	404.1±61.9	313.0±21.5	0.8
January 1965 (1)	156.4	3.71	2.40	124.4	31.6	349.1	318.5	0.7
February 1965 (5)	161.6±12.4	3.03±0.49	5.32±1.08	123.5±10.2	38.4±14.5	449.3±102.3	331.7±26.1	0.8
March 1965 (5)	156.9±11.0	382±0.42	5.40±1.92	125.9±25.2	34.8±12.6	443.2± 93.3	327.1±32.9	0.6

[a]After Gilles-Baillien and Schoffeniels, 1965.

Baillien (1966) and Gilles-Baillien and Schoffeniels (1968), that at the end of the hibernating period the potassium concentration is the highest in the small intestine while that of sodium is the highest in the colon. Taurine and urea increase while the amino acid concentration decreases during hibernation. These results indicate that from the month of August onward the composition of the blood and of the tissues is subjected to important variations well before the apparition of the behavior characteristics of hibernation (burying and torpidity). It is generally assumed that this behavior is induced by a change in temperature. Thus Gilles-Baillien and Schoffeniels (1968) have kept at 20°C tortoises ready to hibernate with the consequence that they did not bury themselves but were partially dormant. The blood analysis revealed no difference with the values found for an animal hibernating normally. Now placing tortoises in the dark at 0°C at the end of the summer immediately induces the burying and the dormancy of the animals. The influence of the photoperiod has then been analyzed. From the month of July on, tortoises have been subjected in a Phytotron to photoperiods of 8 and 16 hours, respectively. The temperature and the relative humidity were kept constant. As far as the behavior is concerned, the motor activity of the animals in photoperiod of 8 hours rapidly slowed down to a complete torpidity while that of the animals in the other conditions remained unchanged. The animals of the first group did not live more than 2 months. The same experiment was performed at the end of the month of September. The animals were sacrificed after one month and the blood analyzed. The results obtained in the two experimental conditions are given in Table 4-3. It can be seen that the only parameter affected by the photoperiod is the sodium concentration. It is also worth mentioning that by placing abruptly active tortoises at low temperature or in a photoperiod (8 hours) characteristic of the winter, the animals rapidly die. Thus it may be concluded from the above results that a lowering of the ambiant temperature induces the burying of the animal while the decrease of the photoperiod is responsible for the dormancy. It is however clear that in the hibernating animal a biological clock must also exist that is responsible for the metabolic changes that make it respond to the solicitations of the environment. This conclusion is drawn from

TABLE 4-3
BLOOD ANALYSIS PERFORMED ON TORTOISES SUBJECTED TO
PHOTOPERIODS OF 8 AND 16 HOURS[a,b]

Group	Photo-period (hours)	Na (mEq/l)	K (mEq/l)	Ca (mEq/l)	Cl (mEq/l)	Urea (mM/l)	Osmotic pressure (mOsm/l)
I[c]	8	145.2	2.74	3.80	77.7	8.17	270.8
	16	125.3	5.45	3.80	89.3	0.86	264.4
II[d]	8	174.1	4.91	4.69	97.7	11.61	285.3
	16	148.3	4.75	5.45	91.3	11.18	314.6

[a]After Gilles-Baillien, 1966 and Gilles-Baillien and Schoffeniels, 1968.
[b]T° and relative humidity are constant.
[c]Group I experiment started at the end of the month of July.
[d]Group II experiment started at the end of the month of September.

the fact that the animal must be metabolically prepared to support adequately the changes in temperature and photoperiod as demonstrated by the fact that an aroused animal dies if abruptly placed in the environmental conditions of hibernation.

Adaptive changes do indeed occur at the level of the activity and distribution of isozymes of lactic dehydrogenase (LDH) in the tissue of a hibernator, as shown by the results of Burlington and Sampson (1968). They have determined the total enzyme activity and the electrophoretic pattern of lactic dehydrogenase in heart, brain, liver, and skeletal muscle from normothermic and hibernating ground squirrels (*Citellus tridecemlineatus*), and normothermic albino rats. As is well known, Kaplan and Cahn (1962) have demonstrated the existence of five isomers in the LDH molecule, each isomer being a tetrameric association of two genetically regulated subunits, heart (H) and muscle (M) types. The distribution of LDH isozymes in various organs has been correlated with metabolic functions: a high proportion of the M form is found in organs that may function anaerobically, muscle for instance, while the H form is predominant in tissues in which aerobic metabolism is predominant such as heart and brain (Dawson *et al.*,1964; see also Chapter 7).

During hibernation, it is found that the M form increases significantly in the heart of the ground squirrels, supporting the idea that the capacity for anaerobic glycolysis is also increased. Compared to tissues from normothermic squirrels, total and M-type liver LDH activity measured at 32° or 15°C increases significantly in the torpid animal. This could reflect a better capacity for the hibernator to direct lactate into the gluconeogenic pathway (Burlington and Klain, 1967). Since the enzyme synthesis is probably not increased during hibernation due to the unfavorable effect of the low temperature on the protein synthesis (Manasek *et al.*, 1965), it may be proposed that the change in distribution pattern of the isozymes results from cytoplasmic factors fluctuating during the life cycle of the species (Burlington and Sampson, 1968). As the ionic composition influences the activity of many dehydrogenases and more specifically LDH and glutamic dehydrogenase (Schoffeniels, 1969), it may be appropriate to look for such an effect in the case of hibernators. This suggestion has a rational basis if one considers Tables 4-2 and 4-4 which show that the ionic composition of the intestinal epithelium and the blood varies during the hibernation of the Greek tortoise.

Another argument favoring this view is that the thermal denaturation of an enzyme is profoundly affected by the ionic composition of the medium. The results of Fig. 4-1 have been obtained with the glutamic dehydrogenase extracted from lobster muscle. The enzyme activity is estimated by measuring the rate of oxidation of the reduced coenzyme as described elsewhere (Schoffeniels, 1966). When heated at 52°C the enzyme is rapidly inactivated. However in the presence of NaCl the effect of heat is still more pronounced (Fig. 4-1), indicating some configuration change under the influence of inorganic ions—a conclusion arrived at by studying the kinetics of the catalytic process in the presence of different salts (Schoffeniels, 1966).

It is clear from the above results that it is difficult to draw a simple generalized conclusion as to the metabolic pattern an organism establishes under various conditions of environmental temperature. The M:T responses, as measured by the oxygen consumption, represent too crude a method to allow a thorough analysis of the phenomenon. The respiration at the level of the whole animal or considered on isolated tissue represents the in-

TABLE 4-4
COMPOSITION OF THE INTRACELLULAR FLUID OF THE INTESTINAL EPITHELIUM
OF THE GREEK TORTOISE *Testudo hermanni* GMELIN
IN AROUSED AND TORPID CONDITIONS[a]

Intestinal epithelium	Na Act. aroused	Na Hib. torpid	K Act. aroused	K Hib. torpid	Ca Act. aroused	Ca Hib. torpid	Amino acids Aroused	Amino acids Torpid
Duodenum	43.0	40.4	68.1	105.5	6.5	3.4	–	–
Jejunum	53.7	43.1	63.2	98.9	8.0	3.7	13.6	4.3
Ileum	56.6	47.9	59.4	90.2	8.1	5.3	–	–
Colon	57.0	75.1	45.1	55.8	10.0	5.1	–	–

[a]Measured in mEq or mM/kg intracellular water. The extracellular space is determined by the inulin method. After Gilles-Baillien and Schoffeniels, 1968.

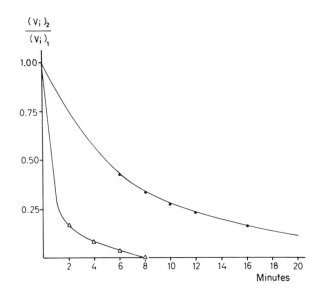

FIG. 4-1. Thermal denaturation of glutamic dehydrogenase extracted from lobster muscle. The results are expressed as a function of time of heating at 52°C. $(Vi)_1$ is the initial velocity of the reaction in the absence of heating. $(Vi)_2$ is the initial velocity observed after heating the enzyme for the time indicated. ● enzyme heated in distilled water. Δ enzyme heated in the presence of 400 mM NaCl. (Gilles and Schoffeniels, unpublished.)

tegration of a large variety of biochemical events, the rate of which might be influenced by other environmental factors that may prove difficult to pinpoint. On the other hand, certain reactions controlling the efficiency of a metabolic sequence may have a more significant influence at room temperature than at others. Accordingly more information concerning the effect of thermal energy on enzyme stability, rate of key reactions, etc. are needed before drawing any conclusion as to the nature of the adaptive mechanism that may prevail in latitudinally separated populations.

In the very active phenomena of sustained flight in some insects, for instance Diptera and Hymenoptera, carbohydrate is the only fuel used-and is supplied to the muscles in the form of trehalose, which appears to represent an immediately available reserve which is used for a very marked increase in respiration when the animal attains full flight.

This appears to be the result of the acquisition of a lateral extension to the glycosis chain in which the reduced NAD generated in the usual system is used to reduce dihydroxyacetone phosphate by the action of an NAD-dependent cytoplasmic 3-glycerophosphate dehydrogenase. The 3-glycerophosphate diffuses into the mitochondrial compartment, where mitochondrial-bound 3-glycerophosphate dehydrogenase is located, and enters directly into the respiratory chain. The appearance, in insects, of an unusual end product of glycolysis, 3-glycerophosphate, which is linked with the adoption of a certain mode of rapid flight in bees and flies, is due to the relative lack of lactic dehydrogenase. The fact that the lactic dehydrogenase is relatively deficient should lead to an accumulation of NADH and, consequently, to a deficit in the NAD necessary for the oxidation of glyceraldehyde 3-phosphate. But the increased concentration of 3-glycerophosphate dehydrogenase prevents this effect. It reduces dihydroxyacetone phosphate to 3-glycerophosphate while oxidizing NADH to NAD. The result is an accumulation of 3-glycerophosphate, which accounts not only for the 3-glycerophosphate cycle active in sustained flight of bees and flies, but also leads to another adaptive radiation of ecological importance: the diapause of nymphs (Schneiderman and Williams, 1953) or of eggs.

Andrewartha (1952) has reviewed the biological aspects of the insect adaptation called diapause. Most commonly, and in insects adapted to the cool temperate zone of the Northern Hemisphere, diapause occurs in the stage of metamorphosis which coincides with the lower temperatures existing during the succession of seasonal changes. During the nymphal diapause undergone by a number of Lepidoptera, the respiration intake falls very markedly to 1/50 of the prediapausal level, and no longer responds to carbon monoxide, cyanide, and azide, all substances which inhibit cytochrome oxidase. This latter effect is due to a lowering of the cytochrome c concentration, from which results a high excess of cytochrome oxidase relative to cytochrome c. In this situation, the low level of cytochrome c causes a fresh formation of reduced NAD, which is reoxidized by an 3-glycerophosphate dehydrogenase, the resulting 3-glycerophosphate being dephosphorylated into glycerol in the tissues. This glycerol protects the insect tissues from the low temperature effects. Salt (1959) has shown that in the sawfly *Bracon cephi* the concentration of glycerol in the larva reaches values as high as 25% of fresh tissues. In these conditions, the larva withstands temperatures as low as −40°C. In the nymphal diapause of *Hyalophora (Samia) cecropia*, glycerol may reach a concentration of 3.5% in the hemolymph. The glycerol comes from glycogen through trehalose. It appears on the day of nymphosis and increases until it reaches a maximum after one or two months (Wyatt and Meyer, 1959; Wilhelm *et al.*, 1961).

In another saturniid, *Philosamia (Samia) cynthia,* the diapausing nymphs accumulate less glycerol, but if exposed to lower temperatures, the glycerol increases in the hemolymph (Wilhelm, 1960). This is not the only biochemical change which is observed in the blood of hibernating nymphs. Data on free amino acids in larvae and nymphs in diapause are available for *Sphinx ligustri* and *Smerinthus ocellatus* (see Chapter V). Alanine levels appear to be much higher in these diapausing nymphs than in their corresponding larvae — another aspect of protection against the effects of low temperature. This is not the case with the nonhibernating nymphs of *Euproctis chrysorrhoea* nor, except for a short time following nymphosis, with another nonhibernating nymph, *Bombyx mori*. If we compare the series of data on aminoacidemia in nymphs and in caterpillars of

Sphinx ligustri, we come to the conclusion that there is less variability of pattern among the diapausing nymphs than among the caterpillars.

The brilliant work of C. M. Williams (1951 – 1952) has focused attention on *Hyalophora cecropia*, in which a reaction of the organism to the external conditions is found when exposure to a low temperature, exerting an influence on the control of the suspension of nymphal diapause, is followed by exposure to a higher temperature. When diapausing nymphs are kept at different temperatures, a consistent result is observed not only for *Deilephila euphorbiae* but also for *Saturnia pavonia* and *Sphinx ligustri* (Bricteux-Grégoire *et al.*, 1960). At a low temperature, the concentration of alanine rises in the hemolymph and the concentration of total glutamic acid falls; the reverse is observed at a higher temperature.

The end of diapause, due to the circulation of ecdysone released from the prothoracic gland under the influence of the brain hormone, is accompanied by increased respiration and a rise in respiratory pigment turnover and concentration. It is also accompanied by a progressive change of pattern of the aminoacidemia, ranging from a pattern characterized by a high proportion of lysine, arginine, alanine, proline, histidine, and total glutamic acid, to a pattern with two peaks: those of total glutamic acid and tyrosine. The change in the steady state of each amino acid is due to new conditions brought about by the circulation of ecdysone and the liberation and incorporation of the amino acids concerned. In the case of glycine, the turnover of the free amino acids from the hemolymph to the adult tissues is higher in nymphs of *Sphinx ligustri* during their progress toward adulthood than in nymphs still in diapause (Florkin, 1966).

In homeotherms, metabolic energy is used to maintain the internal medium at a more or less constant temperature. Heat liberated by exothermic reactions thus limits any ecobiochemical hazard. Here again we find adaptations working at various levels of organization: from the anatomical adaptation provided by the structure of the vascular system in the extremities of cold water mammals (thermic exchanger) down to the molecular aspects of vasoconstriction, shivering, modified intracellular pathways of hydrogen transport, etc. When we remember that the free energy liberated by the reduction of oxygen is of the order of 52 kcal M^{-1} and is partially recovered (to the extent of about 50%) in the respiratory chain as ATP, we have at our disposal a system that may provide quite a lot of energy in the form of heat. The respiratory chain may function even in the absence of phosphorylation and many agents are known that uncouple the synthesis of ATP from the electron transfer. The most important ones, from a physiological point of view, are certainly thyroxine and the long-chain unsaturated fatty acids.

As is well known there are three sites of phosphorylation (I, II, and III) along the respiratory chain and according to the type of substrate used the reducing equivalents may enter the chain at various levels with the consequence that the efficiency, in terms of oxidative phosphorylation, may be reduced. This efficiency is characterized by the so-called P:O ratio and it is maximal, i.e., 3, when 3 inorganic phosphate molecules are used to synthesize 3 ATP from ADP for each pair of electrons transferred to an oxygen atom.

As stated above, the efficiency may be reduced down to one if the two first phosphorylating sites of the respiratory chain are shunted, for instance if cytochrome c accepts the electrons from NADPH instead of NADH. The energy that could have been used at sites I and II to synthesize ATP is released as heat (Fig. 4-2). Thus by shifting from NADH- to NADPH-dependent pathways, a decrease in net phosphorylative efficiency occurs with an enhancement of the heat production.

Under the conditions of cold-stress, an homeotherm shows not only a marked increase in the oxygen consumption but also in the thyroxine production. Thyroxine not only uncouples the oxidative phosphorylation but also inhibits the enzyme (trans-hydrogenase) that is responsible for the reduction of NAD from NADPH. The "caloric shunt" is accordingly more active since much of the free energy liberated by the electrons along the carriers is evolved as heat.

The molecular aspects of cold adaptation in homeotherms appears therefore to depend on the existence of two main pathways of intracellular hydrogen transport. Thyroxine is certainly an important means of control but the role of extramitochondrial cytoplasmic factors in the regulation of electron transport has

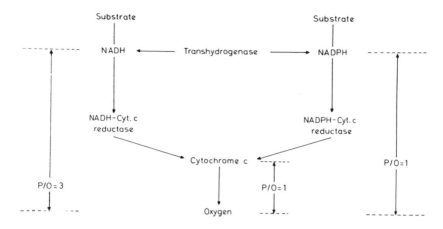

FIG. 4-2. Simplified version of the respiratory chain showing that by shifting from NADH- to NADPH-dependent pathways, a decrease in net phosphorylative efficiency results in an enhancement of the heat production. The transhydrogenase is inhibited by thyroxine. (After Smith, 1960.)

still to be appreciated. The way the cell controls the transfer of reducing equivalents from the cytoplasm into the mitochondria is little known. Results favoring the idea that the "shuttle" systems are partly dependent on the ionic composition of the cytoplasm are presented in Chapter VII.

The fact that the heat production in homeotherm, the production of glycerol in hibernating insects, the adaptation to sustained flight in bees and flies, and the synthesis of amino acids in euryhaline species (see Chapter VII) depend on the existence of means of controlling the intracellular pathway of hydrogen transport offers a new example of physiological radiation of a biochemical system.

An interesting aspect of biochemical adaptation is provided by the consideration of some properties of collagen fibers. The protein collagen plays an important role in controlling the properties of skin structure. The collagen molecule is composed of three polypeptide strands, cross-linked through hydrogen bonding. The hydroxyl group on the hydroxyproline residues present in the frequent sequence gly-pro-hypro is a major contributor to the formation of such secondary bonds. The pyrrolidine residues (pro + hypro) appear, because of steric configuration they impose upon the peptide chain, as important factors in determining and stabilizing the helix configuration of the molecule. This is well demonstrated by considering the temperature at which the helix-random coil transition occurs. As shown by the results of Table 4-5, there is a direct relation between the total pyrrolidine content of various collagen molecules and their thermal transition temperature.

TABLE 4-5

ESTIMATION OF THE THERMAL TRANSITION TEMPERATURE T_m (HELIX→COIL) FOR VARIOUS COLLAGENS[a]

Collagen	Total pyrrolidine content (residues/1000 residues)	T_m (°K)
Cod skin	152	285
Cod swim bladder	160	289
Dogfish sharkskin	166	289
Earthworm cuticle	173	295
Carp swim bladder	193	302
Perch swim bladder	199	304
Rat skin	214	309
Calf skin	232	312
Ascaris cuticle	310	325

[a]After Bailey, 1968.

An outstanding feature of the collagen fiber is the sharp contraction on heating to about one-third of its initial length. Shrinkage temperature (T_S) of mammalian collagens is around $62-65°C$ but among the fish collagens considerable variations occur and range from 38 to 54°C (Table 4-6). The value of T_S may vary with the pH and the composition of the solution (electrolytes, nonelectrolytes), and it is now generally accepted that the shrinkage of the fibers is due to the breakdown of the crystalline regions associated with the pyrrolidine residues (Fig. 4-3). A comprehensive survey of the collagen properties in relation to their chemical structure has recently been presented by Bailey

TABLE 4-6

EFFECT OF PYRROLIDINE RESIDUE CONTENT ON THE THERMAL STABILITY OF VARIOUS COLLAGENS[a]

Source of collagen	Total pyrrolidine content (residues/1000)	Proline content (residues/1000)	Hydroxy-proline content (residues/1000)	Shrinkage temperature (T_S) (°C)	Denaturation temperature (T_D) (°C)	$T_S - T_D$ (°C)	Approximate environmental temperature (°C)
Mammals							
Rat skin	222	130	92	63	37	26	37
Calf skin	232	138	94	65	37	28	37
Aquatic mammals							
Whale	220	128	92				37
Warm water fish							
Shark skin	191	113	78	53	29	24	
Pike skin	199	129	70	55	27	28	24–28
Carp skin	197	116	81	54	29	25	
Cold water fish							
Cod skin	155	102	53	40	16	24	
Dogfish skin	156	99	57	41	16	25	10–14
Invertebrates							
Ascaris cuticle	310	291	19	59	52	7	37
Lumbricus cuticle	173	7.7	165	44–46	22	24	–

[a]After Bailey, 1968.

63

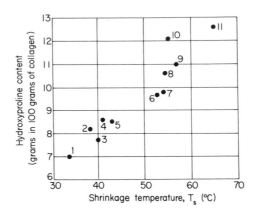

Fɪɢ. 4-3.   The relationship between the hydroxyproline contents of the collagens of various fishes and their "shrinkage temperature." Points 1 to 5: cold-
water fishes; points 6 to 11: warm-water fishes. (Anfinsen, 1959 from results of
Takahashi, cited by Gustavson, 1956.)

(1968). Takahashi and Gustavson (1956) have established a correlation between T_S and the temperature of the habitat. The
skins of deep or cold-water fish, e.g., cod, have a low T_s, around
$33-37°C$, while warm-water fish such as shark and goldfish have
higher values, in the range of $52-56°C$. The denaturation temperature defined as the midpoint of the transition is also correlated with the proline and hydroxyproline content of the collagens. Since the denaturation involves the splitting of the triple
helix, the denaturation temperature is therefore dependent on
two factors: the total pyrrolidine content and the interchain hydrogen bonding. It is interesting to notice that the difference
between the shrinkage and denaturation temperatures is rather
constant.

The great difference in T_S and T_D found when considering collagens of different origin is thus explained on the basis of their
difference in chemistry. The most remarkable feature however is
that some correlation may be established with the environmental temperature, thus pointing to some of the adaptive changes

that affect a molecule controlling important properties of skin structure.

Leaving the animal world, we may consider the bacteria which are able to live at high temperatures. In some of these, there are indications that the proteins have greater resistance to the denaturing influence of heat, but this is not a general characteristic of thermophilic bacteria. As shown by Allen (1950, 1953), the proteins of eurythermal thermophilic bacteria are not particularly thermostable. These bacteria are nevertheless able to grow at temperatures as high as 60–65°C. It appears that damage is done to the cells growing at high temperatures, as well as it would be to the nonthermophilic forms, in the same conditions. But the thermophilic forms have a more active metabolism and repair the cells continuously. A consequence of this fact is that the chemical needs of the thermophilic bacteria become, in some cases, as shown by Campbell and Williams (1953), nutritionally more exacting when the temperature is increased. As shown in Table 4-7, *Bacillus stearothermophilus* 3084 requires biotin and folic acids when grown at 36°C, but as a supplement it needs thiamine if grown at 55°C. These variations of chemical needs are of great ecological interest.

Light

The domain of photobiology is limited to a very narrow band in the electromagnetic spectrum, between 380 and 760 millimicrons. Wald (1959) has emphasized the concept of the fitness of this narrow range in relation to life, i.e., its ecological fitness. As a matter of fact, radiations below 300 millimicrons denature proteins and depolymerize nucleic acids. Terrestrial life is possible only because the earth is surrounded by an ozone layer which forms an opaque screen for short-wave radiations. Within the narrow range defined above are situated the vision of animals and man, the bending of plants toward light, and all forms of photosynthesis. The molecular properties of chlorophyll which are of use in photosynthesis in relation to the light in the medium are "a high receptivity to light, an inertness of structure permitting it to store the energy and relay it to other molecules, and a reactive site equipping it to transfer hydrogen in the critical reac-

TABLE 4-7
INFLUENCE OF TEMPERATURE ON THE NUTRITIONAL REQUIREMENTS
OF THERMOPHILIC BACTERIA[a]

| Organism | Nutritional requirements at | | |
	36°C	45°C	55°C
Bacillus coagulans 2	Histidine Thiamine Biotin Folic acid	Same as at 36°	Same as at 36°
Bacillus coagulans 1039	Thiamine Biotin Folic acid	Same as at 36°	Histidine Methionine Thiamine Niacin Biotin Folic acid
Bacillus coagulans 32	Methionine Tryptophan Thiamine Biotin	Histidine Methionine Tryptophan Thiamine Biotin Folic acid	Same as at 45°
Bacillus stearothermo- philus 1356	No growth	Leucine Thiamine Niacin Biotin	Same as at 45°
Bacillus stearothermo- philus 3084	Biotin Folic acid	Same as at 36°	Thiamine Biotin Folic acid
Bacillus stearothermo- philus 4259	Biotin Folic acid	Methionine Histidine Niacin Biotin Folic acid	Same as at 45°
Bacillus stearothermo- philus 5149-5	Methionine Thiamine Biotin Folic acid	Biotin Folic acid	Same as at 45°

[a]According to Campbell and Williams, 1953.

66

tion that ultimately binds hydrogen to carbon in the reduction of carbon dioxide" (Wald, 1959). These favorable properties are nevertheless accompanied by a less favorable aspect, the fact that chlorophyll absorbs best at the ends of the sunlight spectrum. But this deficiency is compensated by the properties of other pigments present: carotenoids in the green and brown algae, and the phycobilins, phycocyanine, and phycoerythrin in the blue-green and red algae. The adaptive nature of the properties of the carotenoids, absorbing light and transferring the energy to chlorophyll in the green and brown algae, is also clear in the function the same pigments play in the phototropism of plants and in the phototropic and phototactic responses of a number of lower invertebrates. There is a relationship of homology between the pigments which absorb the light that stimulates vision in the eyes of arthropods, mollusks, and vertebrates. Carotenoids synthesized by plants and active in their phototropism are consumed by the animals and used by them to obtain the eye pigment. We are confronted here with one of the numerous physiological radiations of carotenoids.

Photosynthesis and phototropism take advantage of the fact that carotenoids can absorb light and transfer the electromagnetic energy to chlorophyll or to an enzyme, while the property of the pigments used in the physiological radiation called vision is its capacity to undergo cis-trans isomerization. The visual pigments are chromoproteins that bleach rapidly when exposed to light. This process is normally counterbalanced in the living eye by regeneration. There are two main series of naturally occurring visual pigments: rhodopsins and porphyropsins. The structure of the chromophorically active moiety has been elucidated by showing that under the action of light retinol (vitamin A) is released from rhodopsin and 3-dehydroretinol (vitamin A_2) from porphyropsin. The extraction of bleached retinas or the destruction of the visual pigment in darkness with chloroform yields another substance, retinaldehyde (retinene 1) in the case of rhodopsin or 3-dehydroretinaldehyde (retinene 2) in the case of porphyropsin.

The reactions, observed by Wald, may be summarized as follows:

Retinol
(vitamin A_1)

Retinaldehyde
(retinene 1)

3- Dehydroretinol
(vitamin A_2)

3- Dehydroretinaldehyde
(retinene 2)

$$\text{Rhodopsin} \xrightarrow{\text{light}} C_{19}H_{27}CHO + \text{lipoprotein} \xrightarrow{\text{dark}} C_{19}H_{27}CH_2OH + \text{lipoprotein}$$

retinaldehyde   opsin   retinol   opsin

Like the carotenoids, retinol and its aldehyde may exist in a number of different configurations due to the numerous double bonds in the polyene chain (cis-trans isomerism).

There exist many rhodopsins and porphyropsins distinguished by the spectral locations of their α-bonds (Bridges, 1965a,b; Dartnall, 1962; Crescitelli, 1958a,b; Morton and Pitt, 1958). For rhodopsins the absorption maxima range from 433 and 440 mμ in honeybee and frog (Goldsmith, 1958a,b; Denton and Wyllie, 1955; Donner and Reuter, 1962; Bridges, 1967), to 562 mμ in chicken (Wald *et al.* 1955), while that of porphyropsins occur over a more restricted range, from 510 mμ in some labrid fishes (see Wald, 1960) to 543 mμ in the fish *Osmerus* (Bridges, 1965a).

The wide variety of pigments found in fishes is in relation to the adaptation to the diversity of spectral light distribution in their aquatic habitats. Fishes living in deep marine water where solar radiation is limited to a narrow band in the blue-green at 480 mμ have rhodopsins absorbing in the same spectral region. Fresh water fishes live in a medium where there is a high proportion of red light, and their pigments have absorption peaks displaced toward the red. Fishes living in intermediate habitats as well as the land vertebrates usually have pigments absorbing near 500 mμ, corresponding to the wavelength of the solar radiation maximum at the earth's surface. Porphyropsins are generally found in freshwater forms, in accordance with the nature of their photic environment. They may be the only visual pigment as is the case for *Cyprinus* (Crescitelli and Dartnall, 1954), for *Lepisosteus* (Bridges, 1964, 1965b), and for *Necturus* (Crescitelli, 1958b). But many freshwater fishes have a mixture of visual pigments and an interconversion between rhodopsin and porphyropsin may be observed in answer to changes of light intensity or to other modifications of environmental factors. Temperate zone freshwater fishes, with a mixture of visual pigments, increase the proportion of rhodopsin in summer. This has for instance been demonstrated in the Cyprinid fish *Notemigonus crysoleucas bascii* (Bridges, 1965c). During the change of different light habitats in the metamorphosis of amphibians from aquatic tadpole to terrestrial adult, a porphyropsin-rhodopsin interconversion also takes place (Wald, 1946, 1960). Wilt (1959) has proposed a direct role of thyroxine in this phenomenon. Naito and Wilt (1962) using tritiated retinol have shown that it can be a precursor of 3-dehydroretinaldehyde. By injecting the labeled compound in the isolated surviving eye of the sunfish, *Lepomis* sp., a freshwater species characterized by a mixture of visual pigments in which porphyropsin predominates, they were able to locate radioactive 3-dehydroretinaldehyde. They also showed that the introduction of thyroid hormone in the eye reduces the rate of transformation.

Whether the conversion takes place between retinol and 3-dehydroretinol or between retinaldehyde and 3-dehydroretinaldehyde is not elucidated. On the other hand the liver is unable, under the conditions of organ culture, to convert retinol to 3-dehydroretinol. These experiments, combined with the results obtained by Wilt (1959), suggest that the bullfrog tadpole converts retinol to 3-dehydroretinaldehyde, in the eye, with a formation of porphyropsin. The thyroid hormone would reduce or abolish this formation and favor the utilization of retinaldehyde to form rhodopsin.

Lampreys, eels, and salmon migrating between fresh- and seawater, also exhibit porphyropsin-rhodopsin interconversions. Salmon and trout (different species of the genera *Oncorrhynchus* and *Salmo*) have mixtures of retinaldehyde and 3-dehydroretin-aldehyde (Wald, 1941; Kampa, 1953; Bridges, 1956; Munz and Beatty, 1965). Beatty (1966) has shown that in salmon, the two visual pigments are not always present in the same proportion and that the predominant one is not always porphyropsin. He observed that in the spawning migration of adult salmon there was a progressive increase in the percentage of a porphyropsin absorbing at 527 mμ in the retina, and this increase (except for the sockeye salmon) resulted in a conversion from a retina having predominantly a rhodopsin absorbing at 503 mμ to one with a preponderance of porphyropsin (526 mμ).

Some relationship between photic environment and spectral sensitivity of visual pigments may also be found in Decapoda. In *Homarus vulgaris*, the spectral sensitivity of the eye is greatest between 516 and 531 mμ, a range "very similar to the spectral composition of attenuated daylight in the water and depths where lobsters live" (Kampa *et al*, 1963). The eye of the hermit crab *Eupagurus bernhardus* L. has its maximum sensitivity near 500 mμ (Stieve, 1960), a property well adapted to the photic environment provided with shallow coastal waters.

In conclusion, it may be stated, in the words of Munz (1965), as follows:

Several investigators have concluded that the visual pigments of deep-sea animals (teleosts, crustaceans and elasmobranchs) have become adapted to the predominantly blue light in that habitat by shifts in λ_{max} to wavelengths below 500 mμ. This type of evolutionary response to the photic environment has occurred in other marine habitats. Some pelagic fishes living near the surface also have so-called 'deep-sea rhodopsins,' well suited to the predominantly blue light in this environment. In turbid, greenish, or yellowish coastal waters, fishes have 'rhodopsins' with λ_{max} above 500 mμ. These ecological groupings of fishes, with visual pigments correlated to the spectral distribution of sunlight in each habitat, also appear to be paralleled by the crustaceans.

In fresh water, both fishes and crustaceans have visual pigments adapted to the predominance of long wavelengths in the photic environment. In fishes, this is accomplished by the retinene 2 pigments, often mixed with the corresponding retinene 1 pigment. Mixtures of rhodopsin and porphyropsin, in changeable proportions, may be adapted to the highly variable nature of the photic environment in many freshwater habitats.

Such broad generalization is nevertheless not without exceptions, for instance, the existence of rhodopsins with absorption peaks above 520 mμ in terrestrial geckos (Crescitelli, 1956, 1960). Furthermore, if the relation between the spectral properties of the visual pigments of vertebrates and of invertebrates is clear in many cases, it is not always so and we must await a greater knowledge of the cone pigments, so far little known in vertebrates, in order to draw a complete correlation between pigment distribution and environmental factors.

It must be deplored that our knowledge of the phototactic response of animals, sometimes improperly called phototropism, still has so little molecular content. Many animals show positive phototaxis at low light intensities and negative phototaxis at higher intensities. This has been shown to obtain in the oligochaete *Allolobophora*, in the polychaete worms *Serpula* and *Spirographis*, in the nauplii of the barnacle *Balanus perforatus*, and in different insects. This results in a frequent gathering of small organisms near the edge of a region of shadow. On the other hand, there are differences among the individuals within a given population, some *Balanus nauplii*, for instance, remaining positively phototactic at an intensity which is above the threshold of negative phototaxis for other individuals. But, in the same case, the positively phototactic individuals, when kept at the same intensity of light, become negatively phototactic. Though the biochemical explanation of these ecological aspects has not yet been provided, the parallelism with the data on the phototropic curvature of *Avena* is striking. Here the activity of an enzyme in auxin production-transport is changed by light. The light is absorbed by a carotenoid pigment intimately associated with the enzyme, or capable of direct interaction with it. If the intensity is increased, a negative curvature takes place as a result of an irreversible bleaching of the photoreceptor pigment at high intensities, before reaction with the enzyme (see Thimann and Curry, 1960).

Another interesting aspect is the fact that phototaxis, in animals, and more particularly the change of *sign* (positive or negative) is influenced by factors other than light. Dilution of the seawater makes certain positive animals become negative, and negative animals more so. On the contrary, concentration of sea-water weakens positive reactions and strengthens negative ones. Many other described influences await molecular classification. In plants, the nature of the molecular photoreceptors has been clearly identified in a number of cases by the study of sensitivity to wavelength. Unfortunately these studies are greatly lacking in animals, undoubtedly partly due to the difficulty in producing light beams of equal energy at different wavelengths and on account of the imperfections of color filters. Reviewing the evidence in a critical manner, Thimann and Curry (1960) consider a number of data to be established as follows, with respect to regions of maximum sensitivity in animals.

Yellow to yellow-green	*Balanus* larvae, *Euproctis* caterpillar
Blue-green	*Palaemonetes, Loligo*
503 mμ	Blowfly larva
Blue	*Cyclops, Chydorus, Diaptomus,* and *Bufo* species (both eyes and skin), probably most amphibians
483 or 495 mμ	*Arenicola* larva
483 mμ	*Lumbricus, Phacus*

A general insensitivity to red has generally been noted in animals. Insensitivity of many of them to the ultraviolet radiations is due mainly to the high ultraviolet absorption of the eye lens proteins. Molecular studies on animal phototaxis have been sadly neglected and should be actively stimulated for the benefit of our ecobiochemical knowledge.

References

Allen, M. B. (1950). *J. Gen. Physiol.* **33**, 205.
Allen, M. B. (1953). *Bacteriol. Rev.* **17**, 125.
Andrewartha, H. G. (1952). *Biol. Rev.* **27**, 50.
Anfinsen, C. B. (1959). "The Molecular Basis of Evolution." Wiley, New York.
Arrhenius, S. (1889). *Z. Phys. Chem.* **4**, 226.
Bailey, A. J. (1968). *In* "Comprehensive Biochemistry" (M. Florkin and E. H. Stotz, eds.) Vol. 26, p. 297. Elsevier, Amsterdam.
Beatty, D. D. (1966). *Can. J. Zool.* **44**, 429.
Bricteux-Grégoire, S., Duchâteau-Bosson, Gh., Florkin, M. and Verly, W. G. (1960), *Arch. Intern. Physiol. Biochim.* **68**, 424.

Bridges, C. D. B. (1956). *J. Physiol. London* **134**, 620.

Bridges, C. D. B. (1964). *Nature* **203**, 303.

Bridges, C. D. B. (1965a). *Cold Spring Harbor Symp. Quant. Biol.* **30**, 317.

Bridges, C. D. B. (1965b). *Vision Res.* **5**, 223.

Bridges, C. D. B. (1965c). *Vision Res.* **5**, 239.

Bridges, C. D. B. (1967). *In* "Comprehensive Biochemistry" (M. Florkin and E. H. Stotz, eds.) Vol. 27, p. 31. Elsevier, Amsterdam.

Brown, L. A. (1929). *Am. Naturalist* **63**, 248.

Bullock, T. H. (1955). *Biol. Rev.* **30**, 311.

Burlington, R. F., and Klain, G. J. (1967). *Comp. Biochem. Physiol.* **22**, 701.

Burlington, R. F., and Sampson, J. H. (1968). *Comp. Biochem. Physiol.* **25**, 185.

Campbell, L. L., and Williams, O. B. (1953). *J. Bacteriol.* **65**, 141.

Crescitelli, F. (1956). *J. Gen. Physiol.* **40**, 217.

Crescitelli, F. (1958a). *In* "Photobiology" (*Proc. 19th Ann. Biol. Colloq.*) p. 30. Oregon State College, Corvallis, Oregon.

Crescitelli, F. (1958b). *Ann. N.Y. Acad. Sci.* **74**, 230.

Crescitelli, F. (1960). *Ann. Rev. Physiol.* **22**, 525.

Crescitelli, F., and Dartnall, H. J. A. (1954). *J. Physiol. London* **125**, 607.

Dartnall, H. J. A. (1962). *In* "The Eye" (H. Davson, ed.) Vol. II, p. 323. Academic Press, New York.

Dawson, D. M., Goodfriend, T. L., and Kaplan, N. O. (1964). *Science* **143**, 929.

Denton, E. J., and Wyllie, J. H. (1955). *J. Physiol. London* **127**, 81.

Donner, K. O., and Reuter, T. (1962). *Vision Res.* **2**, 357.

Duchâteau, Gh., and Florkin, M. (1955). *Arch. Intern. Physiol. Biochim.* **63**, 213.

Edwards, G. A. (1946). *J. Cellular Comp. Physiol.* **27**, 53.

Edwards, G. A., and Irving, L. (1943a). *J. Cellular Comp. Physiol.* **21**, 169.

Edwards, G. A., and Irving, L. (1943b). *J. Cellular Comp. Physiol.* **21**, 183.

Fawcett, D. W. and Lyman, C. P. (1954). *J. Physiol. London* **126**, 235.

Florkin, M. (1966), "A Molecular Approach to Phylogeny," Elsevier, Amsterdam.

Fraenkel, G., and Hopf, H. S. (1940). *Biochem. J.* **34**, 1085.

Gilles-Baillien, M. (1966). *Arch. Intern. Physiol. Biochim.* **74**, 328.

Gilles-Baillien, M., and Schoffeniels, E. (1965). *Ann. Soc. Roy. Zool. Belg.* **95**, 75.

Gilles-Baillien, M., and Schoffeniels, E. (1968). In preparation.

Goldsmith, T. H. (1958a). *Ann. N.Y. Acad. Sci.* **74**, 266.

Goldsmith, T. H. (1958b). *Proc. Natl. Acad. Sci. U. S.* **44**, 123.

Gustavson, K. H. (1956). "The Chemistry and Reactivity of Collagen." Academic Press, New York.

Hoar, W. S., and Cottle, M. K. (1952). *Can. J. Zool.* **30**, 49.

Horwitz, B. A., and Nelson, L. (1968). *Comp. Biochem. Physiol.* **24**, 385.

Johnston, P. V., and Roots, B. I. (1964). *Comp. Biochem. Physiol.* **11**, 303.

Kampa, E. M. (1953). *J. Physiol. London.* **119**, 400.

Kampa, E. M. Abbott, B. C., and Boden, B. P. (1963). *J. Marine Biol. Assoc. U.K.* **43**, 683.

Kaplan, N. O., and Cahn, R. D. (1962). *Proc. Natl. Acad. Sci. U.S.* **48**, 2123.

Kinne, O. (1956). *Biol. Zentr.* **75**, 314.

Manasek, F. J., Adelstein, S. J., and Lyman, C. P. (1965). *J. Cellular Comp. Physiol.* **65**, 319.

Mangum, C. P. (1963). *Comp. Biochem. Physiol.* **10**, 335.

Morton, R. A. and Pitt, G. A. J. (1958). "Math. Phys. Lab. Symp. No. 8," p. 109. H.M.S.O., London.

Munz, F. W. (1965). *In* "Ciba Foundation Symposium on Physiology and Experimental Psychology of Colour Vision" (G. E. Wolstenholme and J. Knight, eds.). Churchill, London.

Munz, F. W. , and Beatty, D. D. (1965). *Vision Res.* **5**, 1.

Naito, K., and Wilt, F. H. (1962). *J. Biol. Chem.* **237**, 3060.

Panikkar, N. K. (1940). *Nature* **146**, 366.

Prosser, C. L. (1950). "Comparative Animal Physiology." Saunders, Philadelphia.

Rao, K. P., and Bullock, T. H. (1954). *Am. Naturalist* **88**, 33.

Roberts, J. L. (1957a). *Physiol. Zool.* **30**, 232.

Roberts, J. L. (1957b). *Physiol. Zool.* **30**, 242.

Salt, R. W. (1959). *Can. J. Zool.* **37**, 59.

Schneiderman, H. A., and Williams, C. M. (1953). *Biol. Bull.* **105**, 320.

Schoffeniels, E. (1966). *Arch. Intern. Physiol. Biochim.* **74**, 665.

Schoffeniels, E. (1969). *In* "Structures and Function of Nervous Tissue" (G. H. Bourne, ed.). Vol. II. Academic Press, New York.

Scholander, P. F., Flagg, W., Walters, V., and Irving, L. (1953). *Physiol. Zool.* **26**, 67.

Smith, R. E. (1960). *Federation Proc.* **19**, Suppl. 5, 64.

Spector, W. S. (1956). "Handbook of Biological Data." Saunders, Philadelphia.

Stieve, H. (1960). Z. *Vergleich. Physiol.* **43**, 518.

Takahashi, T., and Gustavson, K. H. (1956). *In* "The Chemistry and Reactivity of Collagens" p. 225. Academic Press, New York,

Tashian, R. E. (1956). *Zoologica*, **41**, 39.

Thimann, K. V., and Curry, G. M. (1960). *In* "Comparative Biochemistry" (M. Florkin and H. S. Mason, eds.) Vol. I, pp. 243–309. Academic Press, New York.

Vernberg, F. J., and Vernberg, W. B. (1966). *Comp. Biochem. Physiol.* **19**, 489.

Wald, G. (1941). *J. Gen. Physiol.* **25**, 235.

Wald, G. (1946). *Harvey Lectures*, **Ser. 41**, 117.

Wald, G. (1959). *Sci. Am.* **201**, 92.

Wald, G. (1960). *In* "Comparative Biochemistry" (M. Florkin and H. S. Mason, eds.) Vol. I, pp. 311–345. Academic Press, New York.

Wald, G., Brown, P. K., and Smith, P. S. (1955). *J. Gen. Physiol.* **38**, 623.

Wilhelm, R. C. (1960). Ph.D. Thesis, Cornell Univ., Ithaca, New York.

Wilhelm, R. C., Schneiderman, H. A., and Daniel, L. J. (1961). *J. Insect Physiol.*, **7**, 273.

Williams, C. M. (1951–1952). *Harvey Lectures* **Ser. 47**, 126.

Wilt, F. H. (1959). *J. Embryol. Exptl. Morphol.* **7**, 556.

Wyatt, G. R., and Meyer, W. L. (1959). *J. Gen. Physiol.* **42**, 1005.

CHAPTER V

Chemical Properties of Organisms Related to Chemical Properties of the Environment

Classically the process of migration of water across a membrane is called osmosis and the pressure that has to be applied to establish equilibrium is the osmotic pressure. Osmosis occurs when there is a gradient of water activity across a membrane.

In a simple physicochemical system, the way the gradient of water activity is built is rather irrelevant, hydrostatic pressure or presence of any type of solutes. In the case of living organisms the situation is completely different and contrary to what has generally been stated, it is *not* the gradient of water activity per se which is responsible for the situation actually found but rather the chemical nature of the solutes that bring the water activity to the observed level. The relations that an aquatic animal establishes with its surroundings are thus dependent on the chemical properties of the environment. Although biologists have known for a long time that differences of composition exist between the external medium, the internal medium, and the intracellular fluid of animals, it is only recently that the molecular mechanisms underlying these relations have been explored. As far back as 1900 the relationships between osmotic pressure, organic constituents, and inorganic constituents have been clearly stated by Léon Fredericq. The classical diagram of Fig. 5-1 illustrates these aspects of cell ecology as well as of the ecology of marine organisms. It is worth remembering how these basic discoveries have been made, as an example of the elegance and ingenuity

CHAPTER V

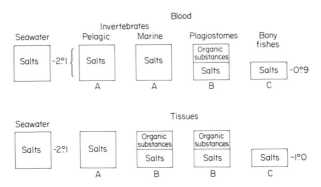

Fig. 5-1.   Osmotic effectors in blood and tissues of animals. (Fredericq, 1901.)

with which the masters of former times were able to solve fundamental biological problems in the simplest of ways. Fredericq,
who was 30 years old at that time, had already discovered the
copper-containing oxygen carrier he had called hemocyanine
and he had published his pioneer work on the distribution of
carbon dioxide between plasma and erythrocytes in mammalian
blood. Fredericq went to the shore of the North Sea and tasted
the blood of a lobster and of other marine invertebrates, and
found it as salty as seawater. He then tasted the blood of a number of sea fishes and found them less salty than seawater, and
about as salty as the bloods of freshwater fishes, themselves more
salty than fresh water. He found that this was also the case with
the blood of freshwater crayfish. Moreover, while cutting the
legs off his crabs in order to obtain the blood he was going to
taste, he discovered the phenomenon he called autotomy, the
mechanism of which he later explained. Fredericq then used
cryoscopy to determine the lowering of freezing point of the different bloods compared with the juice extracted from the tissues
and determined the amount of ash in the two series of samples.
He accounted for the difference in molecular concentration resulting from the cryoscopy experiments and the weighing of the
ash, by postulating the existence of important amounts of small
organic molecules in the tissues of marine invertebrates and in
the blood and tissues of elasmobranch fishes. Urea was identified by Rodier (1900) in the latter case, but the nature of the in-

tracellular organic components of the marine invertebrates, compensating for the lack of high concentration of inorganic constituents despite osmotic equilibrium with seawater, was determined only recently, when in 1951 Camien *et al.* analyzed the free amino acid component of the muscles of various aquatic species. Fifteen amino acids were determined in deproteinized muscle and blood by a microbiological method.[1]

The protein-free samples (tungstic filtrates or dialyzates) were regularly hydrolyzed to avoid ambiguities resulting from the variable activities of "combined amino acids" which might otherwise be present. Higher values were found in the hydrolyzate only in the cases of glutamic and aspartic acids. Asparagine is much less active than aspartic acid (Hac and Snell, 1945). Glutamine itself is as active as glutamic acid (Pollack and Lindner, 1942), but it is converted to inactive pyrrolidone carboxylic acid when autoclaved during the process of analysis. It appears therefore that a part of glutamic and aspartic acids is present as their amides. As shown by Shaw (1958), in the process of the preparation of the samples of crustacean muscles, the amount of extracellular fluid present may be considered as negligible. On the other hand, Table 5-1 shows that the concentrations of amino acids in blood serum of the lobster are much smaller than in the muscles, and column 6 of Table 5-1 excludes the possibility of remaining amounts of extracellular fluid having to modify the conclusions that, in lobster muscle, the nonprotein amino acids are essentially intracellular. Glycine, proline, arginine (including phosphoarginine which is hydrolyzed during the preparation of the samples), glutamic acid (total = free acid + amide), and alanine were the most abundant of the amino acids in lobster muscle, and the molar concentrations of the 15 amino acids determined amounted to 440 mM per liter of intracellular water, while the mM of chloride were 450 per 1000 gm of serum. Therefore, in the lobster, the amino acids determined are important effectors for the intracellular osmolar concentration and their contribution amounts to about half of its value.

The same amino acids account for 360 mOsm per liter water in

[1] In subsequent work a Beckman-Spinco or Technicon automatic analyzer was used.

TABLE 5-1

FREE AMINO ACID CONTENT OF LOBSTER (*Homarus vulgaris*) MUSCLE[a]

Nonprotein amino acids	In muscle per 100 gm of fresh tissue			In blood serum per 100 gm of serum (mg)	In extracellular fluid per 100 gm of fresh muscle (mg)
	Regular (mg)	Dialyzed (mg)	Unhydrolyzed (mg)		
Gly	1025	892	1025	24	3.5
Pro	728	707	728	6	0.9
Arg	778	830	778	1.6	0.2
Glu	267	267	44	3.5	0.5
Ala	133	133	133	8.7	1.3
Asp	12	13	4.6	7.0	1.0
Lys	23	25	24	2.1	0.3
Thr	8.6	8.1	9.1	0.0	0.0
Val	22	22	21	0.0	0.0
Leu	9.3	8.9	8.9	4.2	0.6
His	7.7	8.3	9.7	3.5	0.5
Met	11	12	12	0.0	0.0
Phe	5.4	5.0	5.2	0.2	0.0
Tyr	1.6	1.8	11	3.3	0.5
Total	3031.6	2933.1	2813.5	64.1	9.3

[a]Camien *et al.*, 1951.

the muscles of *Perinereis cultrifera,* 240 in the muscles of *Nereis diversicolor* (Jeuniaux *et al.*, 1961b), 430 in the muscles of *Arenicola marina* (Duchâteau-Bosson *et al.*, 1961), 330 in the gastric caeca of *Asterias rubens* (Jeuniaux *et al.*, 1962), 300 in the muscles of *Leander squilla* (Jeuniaux *et al.*, 1961a), 400 in the muscles of *Carcinus maenas* (Duchâteau *et al.*, 1959), 350 in the muscles of *Eriochéir sinensis* adapted to seawater (Bricteux-Grégoire *et al.*, 1962). The tissues of the species mentioned above are approximately in osmotic equilibrium with seawater (see Chapter VI) and their tissues consequently have an osmotic pressure of about 1000 mOsm per liter. The same amino acids contribute for 170 mOsm in the muscle of *Astacus astacus* (Duchâteau-Bosson and Florkin, 1961), the tissues of which are in equilibrium with a blood containing 430 mOsm per liter.

The existence of lower osmotic contributions of the same amino acids in freshwater forms than in marine forms is shown by Table 5-2. In spite of the existence of the mechanism of osmoregulation in freshwater forms, the concentration of inorganic constituents is always lower in these forms than in the marine ones and the equilibrium between cells and body fluids is maintained by a lowering of the concentration of organic constituents.

While the tissues of invertebrates contain variable amounts of free amino acids they appear to exist at low concentrations in the body fluids, as shown by Tables 5-3 and 5-4. The adaptation to terrestrial life has also led to a lowering of the inorganic constituents of the blood, and the equilibrium between tissues and internal medium has been accomplished through a decrease in the

TABLE 5-2
NONPROTEIN AMINO ACIDS[a]

	Nonprotein amino acid content in:				
Amino acids	Mytilus edulis (marine) (mg/100 gm fiber water)	Ostrea edulis (marine) (mg/100 gm fiber water)	Anodonta cygnea (freshwater) (mg/100 gm fiber water)	Sipunculus nudus (marine) (mg/100 gm fiber water)	Hirudo medicinalis (freshwater) (mg/100 gm fresh tissue)
1. Ala	340	646	8.8	198	8.5
2. Arg	415.5	66.6	36.5	797	1.97
3. Asp	200.4	26.1	4.4	51.8	9.2
4. Glu	317	264	29.4	76.4	81.6
5. Gly	399	248	13.2	>2200	2.1
6. His	12.1	22.9	2.5	6.0	2.6
7. Ileu	24.8	19.2	6.3	13.8	6.4
8. Leu	15.4	12.9	3.6	5.1	5.0
9. Lys	39.4	22.0	8.2	22.0	4
10. Met	9.8	8.4	0.4	4.6	1.6
11. Phe	9.6	8.5	1.6	2.0	5.2
12. Pro	29.0	166	1.0	32.1	7.6
13. Thr	30.5	9.7	3.6	13.6	2.0
14. Tyr	12.7	10.3	2.2	4.6	7.9
15. Val	14.4	10.8	3.3	4.3	11.7

[a]Duchâteau et al., 1952.

TABLE 5-3

AMINO ACID COMPONENT OF MUSCLES AND PLASMA[a]

Amino acid	Rat[b]				Rat[c]		Rabbit[d]		Rabbit[d]				Cat[e]				Rooster[d]		
	Plasma		Muscle[f]		Muscle[f]		Plasma[g]	Muscle[g]	Plasma[f]		Muscle[d]		Plasma[h]		Muscle[h]		Plasma[g]		Muscle[g]
	NH	H	NH	H	NH	H	H	H	NH	H	NH	H	NH	H	NH	H	H	NH	H
1. Ala	—	—	—	—	—	—	6.7	18.0	—	8.2	—	16.9	7.0	7.3	24.7	24.3	4.6	11.1	5.2
2. Arg	2.6	3.6	6.3	4.5	5.6	6.1	0.9	2.9	2.5	2.2	3.5	3.0	1.4	1.6	2.7	3.0	2.2	1.3	1.3
3. Asp	—	—	—	—	—	—	2.1	5.1	0.0	1.7	2.5	7.2	0.1	1.4	3.9	9.8	3.4	2.2	3.2
4. Glu	—	—	—	—	—	—	13.1	73.5	3.4	18.3	19.3	60.9	1.8	3.2	36.2	156.0	16.5	10.7	23.0
5. Gly	—	—	—	—	—	—	9.9	44.3	5.2	4.5	40.9	41.2	2.3	1.7	6.7	17.2	4.0	12.3	12.3
6. His	0.6	1.2	6.2	31.8	8.3	31.0	0.7	31.5	2.1	2.0	3.7	61.7	1.4	1.7	3.6	103.0	0.9	1.6	161.2
7. Ileu	1.0	1.8	3.0	3.0	3.4	3.8	2.4	1.9	2.1	1.5	2.7	2.9	0.8	0.8	1.7	1.6	2.8	0.9	1.2
8. Leu	1.9	5.5	2.6	4.2	5.9	5.9	2.7	2.3	1.9	1.8	3.5	3.6	1.6	1.7	2.3	2.6	4.3	1.6	1.6
9. Lys	3.5	5.6	9.2	9.2	7.2	7.2	2.5	6.8	4.3	4.2	4.4	5.6	2.8	2.8	5.5	9.6	2.1	1.7	2.5
10. Met	0.9	1.7	2.2	1.8	1.1	1.1	0.3	1.3	0.5	0.5	2.2	1.9	0.4	—	0.4	—	1.0	1.1	1.0
11. Phe	2.7	2.8	2.1	2.1	3.8	3.8	1.5	1.7	1.1	1.1	2.1	2.2	0.9	1.0	1.0	1.6	2.3	1.3	1.3
12. Pro	4.3	7.2	7.1	7.4	—	—	4.4	6.9	3.1	3.1	6.8	6.7	2.3	2.0	3.2	8.5	3.5	3.4	4.0
13. Thr	1.8	8.0	14.5	14.0	6.4	7.0	1.6	2.7	1.5	1.6	3.8	4.3	1.4	1.6	3.9	4.6	3.7	2.5	2.8
14. Tyr	2.1	3.1	4.3	3.8	—	1.0	1.0	1.6	0.9	0.6	2.3	2.2	0.7	0.5	0.5	1.1	1.7	1.5	1.6
15. Val	—	4.1	3.2	3.1	5.1	2.8	2.8	2.7	2.8	2.8	4.3	4.3	2.4	2.4	2.3	2.9	4.1	1.5	1.3
Sum 1–15							52.6	203.2		54.1		224.6		40.4		345.8	56.4		223.5
16. Ser													2.1	2.1	5.4	6.1			
17. Cys													<0.2	0.9	0.4	0.1			
18. Try													<0.2	—	<1.05	—			
19. Orn													0.2	0.4	0.4	0.5			

[a] From Florkin, 1956. Measured in mg per 100 gm of fresh tissue or per 100 ml of plasma. NH, dialyzate or monhydrolyzed filtrate; H, hydrolyzed dialyzate or filtrate.
[b] Microbiological method. From Schurr et al., 1950.
[c] Microbiological method. From Solomon et al., 1951.
[d] Microbiological method. From Duchâteau and Florkin, 1954a.
[e] Chromatography on Dowex 50 × 4 according to Moore and Stein. From Tallan et al., 1954.
[f] Tungstic filtrate.
[g] Dialyzate.
[h] Picric filtrate.

TABLE 5-4
BLOOD PLASMA OF INVERTEBRATES, HYDROLYZED DIALYZATES,
CONCENTRATION OF APPARENT AMINO ACIDS (MG PER 100 ML PLASMA)

Amino acids	Anodonta cygnea L.[a]	Parascaris equorum[a]	Homarus vulgaris Milne-Edwards[b]	Astacus fluviatilis L.[c]
1. Ala	0.02	9.5	8.7	10.2
2. Arg	< 0.06	5.1	1.6	3.6
3. Asp	0.09	7.8	7.0	2.3
4. Glu	0.39	21.4	3.5	29.8
5. Gly	0.50	5.7	24.0	6.0
6. His	0.30	0.7	3.5	1.1
7. Ileu	0.09	4.2	–	6.0
8. Leu	0.09	5.7	4.2	3.0
9. Lys	0.15	6.7	2.1	3.0
10. Met	0.01	4.4	0.0	–
11. Phe	0.03	4.4	0.2	1.0
12. Pro	0.03	6.2	6.0	3.3
13. Thr	0.25	6.2	0.0	3.3
14. Tyr	0.04	1.8	3.3	1.7
15. Val	0.08	4.9	0.0	6.0
Total	± 2.13	94.7	64.1	80.3

[a] Duchâteau and Florkin, 1958.
[b] Camien et al., 1951.
[c] Duchâteau and Florkin, 1954b.

amount of small organic molecules present in cells as appears from the values of the same 15 amino acids in the muscles of a number of vertebrates found in Table 5-3. The values of the total of these components: 203 and 224 mg per 100 gm of fresh tissue in the rabbit, 345 in the cat, and 223 in the chicken, contrast strikingly with the value of 3000 in the lobster. This does not however bring in a radical change of cell composition with respect to the inorganic constituents, which appear to be preserved by the adaptive mechanism we have just described.

The intracellular free amino acid pool of invertebrates is not systematically higher in total concentration than that of vertebrates, and, for instance, the total concentration in the muscles of *Hirudo medicinalis* and of *Anodonta* is smaller than in all vertebrates studied so far. There is also in the steady state of the intracellular free amino acid pool a characteristic pattern for each cell differentiation in each organism and therefore "with a grain of salt" a form of taxonomic characteristic. It is however difficult to assess the taxonomic value of a pattern of concentrations of biochemical constituents because the homologous aspects, situated at the level of the enzymes active in the biochemical system involved, are separated from the concentration measured, by a number of intermediary steps. In the case of the free amino acid pool in a cell, the amount and nature of the amino acids present depends on a number of metabolic features: protein or polypeptide synthesis, protein breakdown, amino acid absorption from the body fluids, amino acid excretion in the body fluids, amino acid biosynthesis by transamination or by other pathways, amino acid breakdown after deamination, use of amino acids for the synthesis of other nitrogen components, etc. But it appears that, in a definite set of conditions, the pattern of the intracellular amino acid pool is characteristic of a given cell differentiation of a given species (Duchâteau and Florkin, 1954b). This is in contrast with the fact, shown by several investigators, that the amino acid composition of the whole protein component of muscle, for instance, is not very different from one species to another (Beach *et al.*, 1943; Block, 1951).

In his studies on hormonal effects on nitrogen metabolism, Lotspeich (1950) has reached the conclusion that the speed of utilization of each free amino acid of blood, in the course of an experimental increase of protein synthesis, is directly proportional in the dog to the relative concentration of this amino acid in the most important protein mass in the organism, i.e., in the muscle. Solomon *et al.* (1951) have compared the amino acid patterns of the proteins of rat muscle and the pattern of the free amino acid pool in the same tissues and have found these patterns to differ markedly.

This conclusion also applies to a number of invertebrates as shown in Fig. 5-2 where the relatively constant pattern of the total protein constituents and the diversity of pattern of the free amino acid pool is illustrated in the cases of *Hirudo medicinalis*, *Homarus vulgaris*, and *Buccinum undatum*. This result is not unexpected. Each cell contains a large number of different proteins, enzymes, or others, different with respect to their composi-

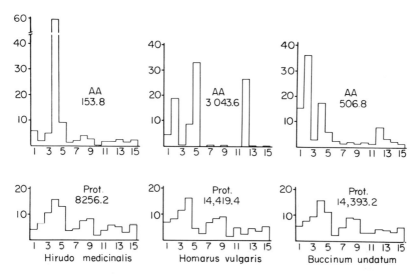

FIG. 5-2.   Pattern of the percentage of the 15 amino acids (numbered 1 to 15 as in Table 5-4) in the free amino acid pool and in the proteins of muscles. (Duchâteau *et al.*, 1954.)

tion and amino acid sequences. The concentrations of these proteins are different and each has a different speed of turnover. In the muscle tissue of the rabbit, for example, the turnover is twice as rapid for aldolase than it is for glyceraldehyde-3-phosphate dehydrogenase (Simpson and Velick, 1954; Heimberg and Velick, 1954). There is nothing astonishing, therefore, in the fact that the total amino acid pattern of a cell can be different from the pattern of the free amino acid pool in the same cell.

As we have seen so far, the difficulty of maintaining, in freshwater and terrestrial species, the concentration of the body fluids at the level of the cell concentration found in marine forms has been solved by the lowering of the intracellular osmotic effect of small organic molecules among which the amino acids play an important role. The ecobiochemical method used has taken a different direction in the insects, where, in spite of a lower inorganic concentration in the hemolymph, the concentration of free amino acids in the cells has been kept at a high level, the osmotic equilibrium with the hemolymph being ensured by a high aminoacidemia.

The free amino acid distribution between tissues and hemolymph of insects has been studied in *Bombyx mori* larvae (Bricteux-Grégoire and Florkin, 1959) and in the southern armyworm, *Prodenia eridania* (Levenbook, 1962). In the silkworm, the dialyzable nitrogen is about twice as concentrated in the intracellular fluid than in the hemolymph water, but the total free amino acid distribution is about the same, despite unequal distribution of certain individual amino acids (Table 5-5). In the southern armyworm, the free amino acid *titres* in fat body intracellular water and hemolymph water are approximately equal. Silkworm and southern armyworm, though both having about the same total concentration of free amino acids in cells and hemolymph, differ with respect to the total glutamic acid (free amino acid + its amide) which is twice more concentrated in the tissues than in

TABLE 5-5
SILKWORM, FIFTH INSTAR[a]

Amino acids[b]	October 1951		July 1953		July 1954	
	Plasma	Tissue	Plasma	Tissue	Plasma	Tissue
1. Ala	45.8	45.0	40.3	58.9	28.4	57.5
2. Arg	37.8	130	32.5	110	33.0	67.1
3. Asp	115	134	105	143	155	96.1
4. Glu	343	612	228	575	236	332
5. Gly	70.1	71.2	60.4	68.6	94.0	70.2
6. His	298	121	178	33.6	252	22.5
7. Ileu	12.7	8.7	11.7	22.5	12.8	12.7
8. Leu	15.8	11.2	12.0	23.9	15.1	15.0
9. Lys	83.9	7.9	69.2	5.2	111	6.4
10. Met	8.0	15.2	10.1	20.0	13.8	7.0
11. Phe	10.5	8.0	10.4	21.1	17.9	11.1
12. Pro	36.6	31.2	19.5	19.1	13.1	12.5
13. Thre	74.3	56.6	62.8	50.5	60.5	34.6
14. Tyr	5.0	12.9	2.4	22.1	3.7	7.1
15. Val	22.9	17.0	19.1	20.4	18.5	14.6
16. Ser			173	121		
Total	1179	1282	1034	1315	1069	766

[a] From Bricteux-Grégoire and Florkin, 1959.

[b] Measured in mg per 100 gm water.

TABLE 5-6
INSECT HEMOLYMPH, HYDROLYZED DIALYZATES, CONCENTRATION OF
APPARENT AMINO ACIDS (MG PER 100 ML PLASMA)[a]

Amino acids	Aeshna sp. larvae, July 1953	Hydrophilus piceus L. adults, January 1951	Apis mellifica L. larvae, October 1951
1. Ala	46.5	60.9	58.4
2. Arg	19.1	7.5	74.8
3. Asp	13.7	18.7	32.4
Free Asp	8.1	—	—
4. Glu	63.3	195.0	308.0
Free Glu	20.0	—	—
5. Gly	54.1	26.1	72.0
6. His	21.3	12.9	30.8
7. Ileu	16.8	25.8	24.8
8. Leu	22.1	7.7	25.0
9. Lys	14.5	24.7	104.8
10. Met	13.7	—	19.6
11. Phe	11.4	7.4	8.8
12. Pro	41.2	283.0	368.0
13. Thre	23.6	17.8	49.6
14. Tyr	13.7	9.2	3.6
15. Val	24.4	20.8	58.4
Total	399.4	717.5	1239.0
Ser	24.4	—	—

[a] Duchâteau and Florkin, 1958.

the hemolymph of the silkworm, while the concentrations are nearly equal in the southern armyworm.

It appears from a general survey of aminoacidemia in insects (Duchâteau and Florkin, 1958) that each species has a particular pattern. Glutamine is more abundant than free glutamic acid, as is the case in mammalian blood, where glutamine is a way of carrying the ammonia of deamination in most tissues. The total concentration of free amino acids is higher in insects than in other animals studied so far. Among insects, the lowest values are found in the representatives of the more primitive orders, those of Odonata (*Aeshna*), Cheleutoptera (*Carausius*), and Orthoptera (*Locusta*). Higher values are found in the more specialized Coleoptera (*Hydrophilus, Leptinotarsa*), Hymenoptera (*Apis*), and Lepidoptera (Table 5-6). In the bee larva as well as in Coleoptera, this higher concentration is mainly accounted for by total glutamic acid and proline. In *Aeshna* a fast of 14 days does not appear to modify the concentrations of the different amino acids (Duchâteau and Florkin, 1958).

The numerous data obtained in Lepidoptera (Table 5-7) concern caterpillars, pupae, and hibernating pupae. If caterpillars and diapausing pupae of *Sphinx ligustri* are compared, one can see that in the caterpillars more than 50% of the total of free amino acids is accounted for by the group glutamic acid (mostly in the form of glutamine)—glycine—histidine (Fig. 5-3). The to-

TABLE 5-7

COMPARISON OF CATERPILLAR AND PUPA IN LEPIDOPTERA,
CONCENTRATION OF APPARENT AMINO ACIDS
(MG PER 100 ML OF HEMOLYMPH PLASMA)

	Euproctis chrysorrhoea L.				*Smerinthus ocellatus*	
	Caterpillars (reared on pear leaves)		Pupae		Caterpillars (hibernation)	Pupae ♀
Amino acids	May 1952	June 1954	June 1952	June 1954	Oct. 1956	April 1955
1. Ala	33.0	—	37.9	—	27.7	171.0
2. Arg	44.8	58.7	99.2	114.7	19.4	319.9
3. Asp	9.0	22.2	24.2	33.7	27.7	1.2
4. Glu	303.2	343.6	279.0	350.5	202.2	127.1
5. Gly	94.3	48.9	63.7	46.3	52.6	38.1
6. His	107.5	161.8	60.5	49.5	83.1	63.6
7. Ileu	15.1	32.9	30.6	45.3	12.5	54.2
8. Leu	13.3	23.6	35.5	43.7	8.3	49.2
9. Lys	50.5	105.3	225.8	103.7	77.6	315.7
10. Met	1.8	13.8	0.0	21.1	8.3	78.8
11. Phe	8.1	15.1	18.5	11.1	9.7	23.7
12. Pro	129.7	157.8	101.6	124.3	23.5	241.5
13. Thr	30.8	54.7	38.7	47.4	34.6	55.1
14. Tyr	0.0	5.3	5.6	20.0	30.1	16.1
15. Val	29.8	49.3	46.0	63.7	83.5	89.8
Total	870.9	1093.0	1066.8	1075.0	700.8	1645.0

[a] From Duchâteau and Florkin, 1958.

Sphinx ligustri

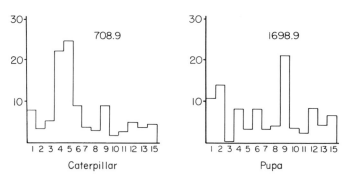

FIG. 5-3. *Sphinx ligustri*, caterpillar and pupa. Figures along the ordinates represent percentages of the sum of the 14 free amino acids considered, the total of which is indicated in each profile (in mg per 100 ml plasma). The figures along the abscissae correspond to the different amino acids indicated in Table 5-4. (Duchâteau and Florkin, 1955.)

tal concentration is higher in the diapausing nymphs, this being due to the higher concentrations of the main amino acids of nymphal hemolymph: arginine, lysine, and proline. The concentration of glycine is lower in nymphs. The same aspect, which cannot be accounted for by a simple dehydration effect, is found in the diapausing nymphs of *Smerinthus ocellatus* compared to the caterpillar. The nondiapausing nymphs of *Euproctis chrysorrhoea*, compared with the corresponding caterpillar, do not show the same kind of difference.

References

Beach, E. F., Munks, B., and Robinson, A. (1943). *J. Biol. Chem.* 148, 431.

Block, R. J. (1951). *Ann. N. Y. Acad. Sci.* 53, 608.

Bricteux-Grégoire, S., and Florkin, M. (1959). *Arch. Intern. Physiol. Biochim.* 67, 29.

Bricteux-Grégoire, S., Duchâteau-Bosson, Gh., Jeuniaux, Ch., and Florkin, M. (1962). *Arch. Intern. Physiol. Biochim.*, 70, 273.

Camien, M. N., Sarlet, H., Duchâteau, Gh., and Florkin, M. (1951). *J. Biol. Chem.* 193, 881.

Duchâteau, Gh., and Florkin, M (1954a). *Arch. Intern. Physiol.* 62, 205.

Duchâteau, Gh., and Florkin, M. (1954b). *Arch. Intern. Physiol.* **62**, 487.

Duchâteau, Gh., and Florkin, M. (1955). *Arch. Intern. Physiol. Biochim.* **63**, 213.

Duchâteau, Gh., and Florkin, M. (1958). *Arch. Intern. Physiol. Biochim.* **66**, 573.

Duchâteau, Gh., Sarlet, H., Camien, M. N., and Florkin, M. (1952). *Arch. Intern. Physiol.* **60**, 124.

Duchâteau, Gh., Florkin, M., and Sarlet, H. (1954). *Arch. Intern. Physiol.* **62**, 512.

Duchâteau, Gh., Florkin, M., and Jeuniaux, Ch. (1959). *Arch. Intern. Physiol. Biochim.* **67**, 489.

Duchâteau-Bosson, Gh., and Florkin, M. (1961). *Comp. Biochem. Physiol.* **3**, 245.

Duchâteau-Bosson, Gh., Jeuniaux, Ch., and Florkin, M. (1961). *Arch. Intern. Physiol. Biochim.* **69**, 30.

Florkin, M. (1956). "Vergleichend biochemischen Fragen. 6. Colloquium der Gesellschaft für physiologie Chemie" pp. 62-99. Springer, Berlin.

Fredericq, L. (1901). *Bull. Classe Sci. Acad. Roy. Belg.* 428.

Hac, L. R., and Snell, E. E. (1945). *J. Biol. Chem.* **159**, 291.

Heimberg, M., and Velick, D. E. (1954). *J. Biol. Chem.* **208**, 725.

Jeuniaux, Ch., Bricteux-Grégoire, S., and Florkin, M. (1961a). *Cahiers Biol. Marine* **2**, 373.

Jeuniaux, Ch., Duchâteau-Bosson, Gh., and Florkin, M. (1961b). *J. Biochem.* **49**, 527.

Jeuniaux, Ch., Bricteux-Grégoire, S., and Florkin, M. (1962). *Cahiers Biol. Marine* **3**, 107.

Levenbook, L. (1962). *J. Insect Physiol.* **8**, 559.

Lotspeich, W. D. (1950). *J. Biol. Chem.* **185**, 221.

Pollack, M. A., and Lindner, M. (1942). *J. Biol. Chem.* **143**, 655.

Rodier, E. (1900). *Compt. Rend.* **131**, 1008.

Schurr, P. E., Thompson, H. T., Henderson, L. M., and Elvehjem, C. A. (1950). *J. Biol. Chem.* **182**, 29.

Shaw, J. (1958). *J. Exptl. Biol.* **35**, 920.

Simpson, M. V., and Velick, S. F. (1954). *J. Biol. Chem.* **208**, 61.

Solomon, J. D., Johnson, C. A., Sheffner, A. L., and Bergeim, O. (1951). *J. Biol. Chem.* **189**, 629.

Tallan, H. H., Moore, S., and Stein, W. H. (1954). *J. Biol. Chem.* **211**, 927.

CHAPTER VI

Isosmotic Intracellular Regulation

The extent to which an aquatic animal is able to maintain constant its cell volume when placed in media of various salinities defined what is generally called the euryhalinity of the species.

The physiologists have mainly been interested in considering the effect of the osmotic pressure of the outside medium on the same parameter, either in the extracellular fluid or the intracellular fluid. When going from a concentrated to a dilute medium the cell is expected, if it behaves as an osmometer, to swell, to reestablish an equality between the activities of water in the solutions separated by the plasma membrane. However the swollen cells may prevent the animal from performing normally elementary functions such as locomotion, feeding, etc. (Lange, 1964), and the true euryhalinity may only be considered to exist if the animal is able to keep its cell volume constant under the conditions of different salinities in the outside medium. Thus there must exist mechanisms that help to keep the water activity within the cell in such a state that the cell volume remains constant despite a change in the activity of the water in the outside medium.

From experimental as well as from theoretical considerations it is obvious that the relation existing between the activites of water inside and outside a cell cannot be maintained by a mere transfer of water across the plasma membrane, thus implying, as the case may be, an increase or decrease in cell volume. It is indeed well known that a polymer, when brought into contact with a suitable solvent, absorbs a large quantity of liquid. Swelling occurs for the same reason that a solvent mixes with a solute to form an ordinary solution. A swollen network acquires an elongated configuration and elastic retraction forces develop in opposition to the swelling process. In the case of ionized polymers such as those found in the cell, the electrically charged groups are screened by the soluble small ions, thus reducing the electrostatic repulsions that may help to expand further the network.

The situation that exists in the case of a swollen ionic network and the surrounding electrolyte closely resembles Gibbs-Donnan membrane equilibria. The polymer acts as its own membrane preventing the charged substituents, which are distributed essentially at random through the gel, much as they would be in an ordinary solution, from diffusing into the outer solution. The swelling force resulting from the presence of these fixed charges may be identified with the net osmotic pressure across the semipermeable membrane in a typical Gibbs-Donnan equilibrium.

It is observed that the concentration of mobile ions will always be greater in the gel than outside because of the attracting power of the fixed charges. Consequently the osmotic pressure of the solution inside will exceed that of the external solution and fluid will be transferred from the solution into the network. The process will eventually come to an end either by rupture and dispersion of the gel, if the swelling pressure exceeds the mechanical resistance of the gel, or by establishment of an equilibrium between the elastic retraction forces and the swelling pressure.

The activity of water inside the gel, however, is difficult to know exactly. The gel lowers the activity of the water in an unknown way as may be deduced from the fact that an osmotic pressure measurement cannot be used to determine molecular weight above 10,000. Moreover the effect of the swelling pressure on the water activity is difficult to ascertain. Finally, as far as solutes are concerned, there is still much uncertainty as to whether we are still in the range where their activity is proportional to their mole fraction.

In the case of living cells, recent progress in biochemistry as well as in electron microscopy favors the idea that the cell interior is highly organized and that the macromolecules, surrounded by the intracellular fluid, form a system closer to the physicochemical state of a gel than a true solution. Inflow of extracellular solution tends to produce a swelling of the cell due to a Gibbs-Donnan effect. For instance, if the red cells were at Gibbs-Donnan equilibrium the osmolarity of the cytoplasm would be expected to exceed that of the medium by 25 to 30 mOsm and the excess hydrostatic pressure to be of the order of 5800 mm H_2O (Tosteson, 1964). This last figure is obviously much too large to be compatible with the mechanical resistance of the cellular membrane. From the value computed by Rand and Burton (1964), we know that the pressure inside the red cell is 2.3±0.8 mm H_2O higher than the outside, a value that agrees rather well with the estimate of Cole (1932) for the internal pressure of a sea-urchin egg. This indicates that the animal cells are not at Gibbs-Donnan equilibrium and that they resort to specialized mechanisms to avoid bursting (Schoffeniels, 1967).

Swelling can be avoided if the cell possesses a rigid membrane or a membrane impermeable to water and to a large fraction of the solutes present in the extracellular medium. The first solution has been adopted by plant cells and some bacteria (Rothstein, 1959). The existence of tough walls that resist the large hydrostatic pressure difference balancing the osmotic inflow of solution thus offers an interesting way of explaining why in the course of evolution the biosynthesis of cellulose and of some mucopeptides did occur. In doing so, the plant cells have had to give up motility in favor of the solution they adopted for the regulation of their cellular volume.

The second solution, i.e., impermeability of the cell, has never been shown to occur. On the contrary, the use of isotopic tracers has demonstrated that the water as well as the inorganic ions contained in a living system can traverse the membrane. Despite these facts in normal conditions of metabolism no swelling of animal cells occurs. But as soon as the metabolic activity of the cell is impaired, cell swelling develops. This phenomenon has been extensively studied by various authors (Opie, 1949; Mudge, 1951; Deyrup, 1953; Leaf, 1956; Whittam, 1956; Robinson, 1960). Some of them consider that the swelling of various tissues occurring in the conditions of metabolic inhibition is indicative of a gradient of water activity between the cell interior and the extracellular medium. However, since the accumulated fluid responsible for the swelling has an activity for water equal to that of the incubating medium and the increase in cell solutes is made up mainly of sodium chloride derived from the external solution, it seems more reasonable to interpret the swelling in terms of a Gibbs-Donnan effect. Additional evidence in favor of a lack of activity gradient for water across the cell membrane may be found in the results of Deyrup (1953) showing that swelling does not occur when the external medium contains sucrose instead of sodium chloride. Thus the presence of a nonpenetrating external solute prevents swelling of animal cells though metabolically impaired.

The extremely high permeability to water of cellular membranes and the smallness of the cells also favor the absence of a gradient of water activity between cytoplasm and extracellular medium. Any gradient of activity that could develop would be rapidly dissipated in considerably less than one second in most cells. Thus water traverses the membrane more rapidly than do virtually all polar solutes and very much more rapidly than do ions. The water activity inside the cell may thus be related to the metabolic activity of the membrane as well as to that of the gell-like network of polymers constituting the integrated metabolic sequences.

It is obvious that any impairment of the metabolic activity of the cells as a whole should lead to a swelling not necessarily because of a difference in osmotic pressure but because of the existence of polymer chains (a) that may accept water as solvent, (b) that contain ionized groups, and (c) that control together with the cellular membrane the activity of the solutes within the cell. In this respect the active transport of cations must be considered, as discussed elsewhere (Schoffeniels, 1967), to be of prime importance in the maintenance of the cell volume.

The cell is thus in a state far from thermodynamical equilibrium and energy must be fed into the system to keep it in this situation. The swelling observed in the conditions of metabolic impairment is thus an illustration of this fact.[1]

[1] The swelling of brain slices incubated in an artificial saline is partially prevented by the addition of a small amount of inulin (Franck *et al.*, 1968). This may indicate that inulin takes the place of water molecules or decreases the elastic properties of some macromolecules present in brain tissue.

There is however some differences according to the type of differentiation considered as demonstrated, for instance, by the ease with which cerebral edema is obtained.

Coming back to the problem that the cells of a euryhaline animal encounter when going from a concentrated to a dilute medium it thus seems inappropriate to consider that we are dealing with a simple physicochemical system that responds to a change in water activity in the outside medium by a parallel change in the intracellular fluid, thus implying a change in cellular volume.

In the case of *Acmaera scutum*, a gastropod found in both marine and estuarine environments, results obtained by Webber and Dehnel (1968) concerning the ion balance indicate that the extracellular volume of foot muscle tissue, as represented by the inulin space, varies from 16.7% in half-diluted seawater to 25.9% in full-strength seawater. This is certainly a reflection of the fact that the muscle fibers are swollen in the dilute medium.

The emphasis has been placed by some authors (Lange, 1964; Lange and Fugelli, 1965; Fugelli, 1967) on this aspect of the problem and, as will be shown below, volume changes have indeed been observed. However the extent or the duration of the swelling (or shrinkage) are inversely related to the euryhaline abilities of a species. Though we do not deny the possible participation of such a phenomenon in the very early stage of the adaptation to media of different salinities, we want to stress the point that the more euryhaline the species is, the less pronounced is this early volume change. Moreover, we consider that the activity of water within the cell is kept in the particular state we observe by the action the metabolic sequences exert on the solute molecules. Furthermore, we do not consider that the difference in the cell composition we observe when a euryhaline animal goes from a concentrated to a dilute medium is a mere result of a physical property, the osmotic pressure, of the environment. We are, on the contrary, of the opinion that the chemical nature of the solutes that bring the water activity of the outside medium to the particular state under consideration is the most important factor in determining the metabolic response of the cell allowing to keep its volume constant. This is well demonstrated by the following experiments (see also Chapter VIII).

When placed in a solution of diluted seawater (23%), speci-

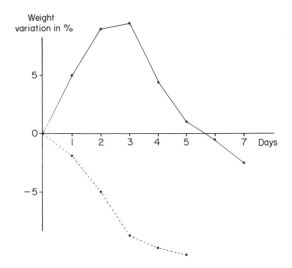

FIG. 6-1. *Rana temporaria* L. Weight variation as percent of initial weight in individuals sitting in diluted seawater (23 %) or in a sucrose solution isosmotic to the diluted seawater. Means of determinations performed on 10 individuals. ●——● seawater 23 %; ●-----● sucrose. (After Schoffeniels and Tercafs, 1965— 1966.)

mens of *Rana temporaria* L. gain weight (Fig. 6-1), lymphatic sacs are swollen, and the liquid that fills them has the same osmotic pressure as the blood. If the animal is placed in a solution of saccharose having the same osmotic pressure, it loses weight as shown by the results of Fig. 6-1 (Schoffeniels and Tercafs, 1965 — 1966).

The contractile vacuole of *Paramoecium caudatum* Ehrenberg plays an important role in the osmo- and ionoregulation of that cell, and it has generally been assumed that the frequency of its formation is controlled by the activity of the intracellular water (Eisenberg-Hamburg, 1925). From results obtained by varying the osmotic pressure of the incubation medium by means of various solutes it may be demonstrated that the frequency of contraction is related to the activity of the inorganic ions and more specifically sodium (Raze and Schoffeniels, 1965). Figure 6-2 shows that in the presence of 28.75 mOsm of NaCl the rate of output decreases while in the presence of sucrose at the same osmolarity it remains unchanged (Fig. 6-3).

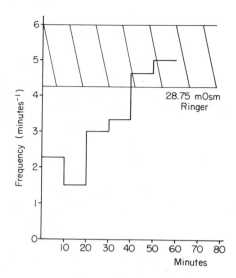

FIG. 6-2. *Paramoecium caudatum* Ehrenberg. Frequency of vacuolar output in dilute frog Ringer. The shaded area corresponds to the limits of variation in the control. (After Raze and Schoffeniels, 1965.)

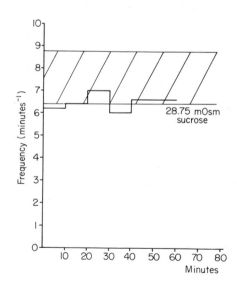

FIG. 6-3. *Paramoecium caudatum* Ehrenberg. Frequency of vacuolar output in a sucrose solution isosmotic to that used in Fig. 6-2. The shaded area corresponds to the limits of variation in the control. (After Raze and Schoffeniels, 1965.)

Some results obtained by Kleinzeller and his collaborators should be mentioned since they point to the possibility of another type of mechanism in the cell volume regulation. Kleinzeller and Knotkova (1964a) have found that, in the steady-state efflux of sodium-24 from kidney cortex slices, ouabain hardly affected the steady-state level of tissue water, while it produced a considerable increase of tissue sodium and a decrease of tissue potassium. Therefore in this case, high levels of tissue sodium and a low content of potassium were not associated with a swelling of the tissue. These authors were also able to demonstrate an ouabain-insensitive extrusion of sodium from sodium-loaded kidney cortex cells as well as a lithium extrusion from lithium-loaded cells. Their results point to a common mechanism responsible for the situation observed with sodium and lithium. Thus, the view that the ouabain-insensitive extrusion of electrolytes from kidney cortex cells is not brought about by a cation pump seems to be acceptable, and one has therefore to turn to another mechanism to explain both the cell volume regulation and the extrusion of electrolytes.

That the ouabain-insensitive electrolyte and water extrusion is due to a mechanical, metabolically dependent force, such as the squeezing of an electrolyte solution from the cell, has been proposed by Kleinzeller and Knotkova (1964a). A contractile mechanism has been proposed to explain the observed volume changes in mitochondria (Lehninger, 1962) and also for some other cells (Wohlfarth-Bottermann, 1963). The difference observed by Kleinzeller and Knotkova (1964a,b) between the rabbit kidney cortex and the rat diaphragm on the one hand and the liver cells where such a phenomenon is not observed, on the other, may be due to the absence of the contractile mechanism in the latter tissue or its damage in the slicing procedure.

With regard to a possible role of some contractile protein in membrane phenomena, the results of Wins and Schoffeniels (1965, 1966a,b, 1967) suggest the identity between a red cell ATPase found to be dependent on calcium ions and inhibited by salyrgan, atebrin, amytal, and 2,4-dinitrophenol, and the erythromyosin described by Ohnishi (1962). A possible model involving the existence of a contractile mechanism in membrane phenomena has been presented elsewhere (Schoffeniels, 1961, p. 638 and 641).

As shown in the preceding chapter, the concentration of the intracellular amino acid pool is higher in marine invertebrates than in those of fresh water—a result that suggests that amino acids are important in lowering the activity of water within the cell and thus playing a significant role in the osmoregulation.

More direct evidence can be found if one studies the variation of the intracellular pool of free amino acids in a euryhaline species, *Eriocheir sinensis* Milne-Edwards for instance, living in media of various concentrations.

Table 6-1 gives the results of analysis performed on the muscle of crabs living in seawater or adapted to fresh water. In both conditions the crabs were fed with fish and the temperature of the water was around 10°C. In these experiments, the inorganic composition of the intracellular fluid has also been determined in order to establish the osmotic balance. Figure 6-4 gives the results obtained with nerve fibers isolated from the meropodites of the claws and walking legs of the same species (Schoffeniels, 1964, 1967). It can be seen that the intracellular composition of the nerve fibers is also subjected to variations. As the hydration of the muscles is approximately the same in seawater and in fresh water (Scholles, 1933), the reversible variation of the amino acid component could only depend on active modification (Florkin, 1956).

As the same kind of difference was found in the concentrations of free amino acids in *Carcinus maenas* living in seawater or in brackish water (Table 6-2), it was proposed to consider the variation of the amino acid component resulting from the change of concentration in the medium, as accomplishing an intracellular regulation "acting against the water movement between cells and body fluids as a consequence of changes of concentration in the latter" (Duchâteau and Florkin, 1956).

In 1958, Shaw studied the phenomenon in *Carcinus maenas*, with the use of fibers of the carpopodite extensor and flexor muscles of the chela. Shaw observed that when the concentration of the external medium is modified (seawater → diluted seawater), the osmotic pressure in the fibers varies proportionally to the

TABLE 6-1
INTRACELLULAR OSMOTIC EFFECTORS IN *Eriocheir sinensis*
ADAPTED TO FRESH WATER AND TO SEAWATER[a]

Intracellular osmotic effectors	Content (mOsm/liter of water) in:			
	Fresh water	Seawater	Fresh water	Seawater
Cl	76.0	153.1	44.6	166.9
Na	68.5	140.8	41.4	146.9
K	56.8	159.0	84.5	133.1
Ca	11.7	8.1	5.2	11.2
Mg	9.2	22.4	9.2	25.3
Total inorganic effectors	222.2	483.4	184.9	483.4
Ala	17.1	46.1	18.1	71.9
Arg	36.7	56.0	36.5	54.7
Asp (total)	5.4	12.2	3.6	11.7
Glu (total)	15.0	36.8	10.3	28.2
Gly	46.5	73.4	57.0	108.5
Ileu	1.4	4.6	1.0	3.2
Leu	2.2	6.1	1.7	5.4
Lys + his + X	9.6	21.7	14.3	18.5
Phe	0.0	tr.	0.0	tr.
Pro	18.2	37.3	4.7	23.7
Ser	5.2	7.6	2.6	6.3
Thr	4.4	17.2	4.4	15.3
Tyr	0.0	tr.	0.0	tr.
Val	0.0	8.1	0.0	6.9
Total amino acids determined	161.7	327.0	154.2	354.3
Tau	14.1	13.6	20.5	27.7
Trimethylamine oxide	49.9	73.9	45.3	75.8
Betaine	9.5	6.9	25.7	21.0
Undetermined nitrogen	108.5	187.7	89.3	131.9
Total effectors determined	565.9	1092.5	520.0	1094.1
Calculated osmolar concentration $(\Delta/1.87) \times 1000$	588.0	1117.6	588.0	1117.6

[a]After Bricteux-Grégoire *et al.*, 1962.

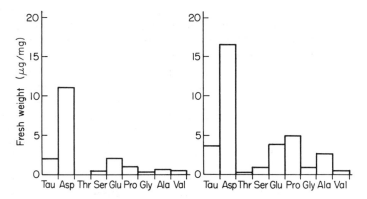

FIG. 6-4.   *Eriocheir sinensis* Milne Edwards. Free amino acids in nerve fibers isolated from animals kept in fresh water (left) and seawater (right); ordinate: μg per mg fresh weight. (After Schoffeniels, 1964, 1967.)

concentration of the body fluids, as a result of the change in the water outside the body. Shaw showed that, when *Carcinus* is transferred from seawater to brackish water, the change in the concentration of the amino acids in the cells is much larger than is accounted for by the slight change in hydration. Shaw also adopts the opinion, according to which the muscle fibers of *Carcinus* are able to limit, by a change of intracellular osmotic pressure, the discrepancy between the cells and the new concentration of body fluids and therefore avoid a bursting out of the tissues as it exists in the stenohaline marine forms when they are introduced into brackish water. The importance of the variation of the total amino acid intracelluar component in the adaptation of *Mytilus edulis* to brackish water was also pointed out by Potts in 1958.

The term "isosmotic intracellular regulation" has been proposed to define the regulation which leads to the active adjustment of the intracellular osmotic pressure to the new osmotic pressure of the body fluids, more or less preventing a change of hydration of the cells (Jeuniaux *et al.*, 1961a). This term replaces by a definite molecular concept the vague notion of "cellular adaptation" and also introduces a distinction between the isos-

TABLE 6-2

Carcinus maenas: AMINO ACID COMPOSITION OF MUSCLES IN
ANIMALS ADAPTED TO SEAWATER OR TO BRACKISH WATER.[a]

Amino acid	Composition (mOsm/kg fresh tissue)		Composition (mOsm/liter of water)	
	Seawater	50/100 seawater	Seawater	50/100 seawater
Ala	19.3	16.0	25.8	20.8
Arg	34.6	32.9	46.3	42.8
Asp	2.5	2.0	3.3	2.6
Glu	31.9	19.9	42.7	25.9
Gly	139.6	75.3	186.9	97.9
His	0.03	0.2	0.4	0.3
Ileu	0.7	0.7	0.9	0.9
Leu	0.8	0.9	1.1	1.2
Lys	1.0	2.5	1.3	3.3
Met	2.0	1.4	2.6	1.8
Phe	0.3	0.3	0.4	0.4
Pro	59.3	28.3	79.4	36.8
Thr	1.0	1.0	1.3	1.3
Tyr	0.2	0.2	0.3	0.3
Val	1.0	1.0	1.3	1.3
Total	294.2	182.6	394.0	237.6

[a]After Duchâteau *et al.*, 1959.

motic intracellular regulation and what is known as the osmotic regulation of body fluids, which is an anisosmotic regulation, keeping the concentration of body fluids above that of the external medium in freshwater forms and below in marine bony fishes.

As we have stated above, the existence of the isosmotic cellular regulation was shown in *Eriocheir sinensis,* the Chinese crab, in 1955 (Duchâteau and Florkin). The phenomenon has been observed repeatedly in many experiments and even when the free amino acids of muscles show a higher concentration in fresh water than that observed in freshly captured animals. This is observed when the animals are kept in aquaria for some time (Duchâteau-Bosson and Florkin, 1962a) and, in this case, the concentration of the intracellular inorganic component is reduced. On the other hand, the phenomenon is also observed in spite of the removal of the eyestalks (Duchâteau-Bosson and Florkin, 1962b). In fresh water, the anisosmotic regulation of the blood keeps the blood concentration measured by freezing point depression (Δ) between 0.99° and 1.2°C (mean 1.10°C), while the Δ of the external water is 0.02°C. When *Eriocheir sinensis* lives in seawater, its blood is in osmotic equilibrium with the external water ($\Delta = -2.09$°C in our experiment) (see also De Leersnijder, 1966, 1967a, b).

In Chinese crabs adapted to fresh water, the inorganic constituents form, as shown in Table 6-3, an important part (40 %) of the osmotically active constituents. The rest of the osmotic pressure is accounted for approximately by the presence of small organic molecules, among which the amino acids determined constitute more than 50 %. The most important of the amino acids, from the osmotic point of view, are glycine, arginine (comprising phosphoarginine which is hydrolyzed during the preparation of the samples), proline, alanine, and glutamic acid (free acid + glutamine). The contribution of trimethylaminoxide to the sum

TABLE 6-3
Eriocheir sinensis: Main Components (mOsm/liter of water) of the Intracellular Isosmotic Regulation[a]

	Inorganic effectors	Amino acids measured	Trimethylamine oxide	Undetermined N	Total non-protein N (Kjeldahl)
In fresh water: ($\Delta_e = 0.02$°C; $\Delta_i = 1.1$°C)	222.2	175.8	49.9	108.5	463.8
In seawater: ($\Delta_e = 2.09$°C; $\Delta_i = 2.09$°C)	483.4	340.6	73.9	187.7	798.6
Change in concentration mOsm/l water	261.2	164.8	24.0	79.2	334.8
Explained by concentration	22.9	18.1	5.6	11.2	47.8
Not explained by concentration	238.3	146.7	18.4	68.0	287.0

[a]After Bricteux Grégoire *et al.*, 1962.

of osmotically active compounds is important, and quantitatively similar to that of glycine. Taurine and betaine contribute to about the same degree as alanine and proline. The contribution of the sum of the undetermined nitrogen constituents amounts to approximately twice that of trimethylaminoxyde or glycine. In spite of the fact that the osmotic coefficient of the constituents determined cannot be considered as equal to unity, it appears that the nitrogenous constituents, added to the inorganic ones, make up the greatest part of the osmotically active components, if not the whole.

When the animals are adapted to seawater, an increase is observed in the concentrations of inorganic constituents, of the amino acids determined, of the trimethylaminoxide and of the undetermined nitrogen. The concentrations of taurine and betaine show no increase. As shown by Table 6-3, the slight reduction of the degree of hydration accounts for only a small part of the change of concentration of the osmotically active constituents. The sum of these osmotically active constituents corresponds to the greatest part, or even the total, of the osmotic component responsible for the new equilibrium which is accomplished between the cell content and the blood plasma, the concentration of which becomes equal to that of seawater. The total concentration of the 15 amino acids determined are approximately doubled when the Chinese crab is transferred from fresh water to seawater, as Duchâteau and Florkin observed in 1955.

This change in concentration is due to changes in most amino acids. Phenylalanine is not present in the free form in the muscles of *Eriocheir sinensis*. The increase of trimethylaminoxide is of the order of 50 %. The total concentration of inorganic constituents is more than doubled. The greatest effect is due to sodium, potassium, and chloride. In the sum of amino acids, alanine, glycine, total glutamic acid, proline, and arginine (including phosphoarginine) contribute with the highest increments, that of trimethylaminoxide being of the same magnitude.

The fact that the pattern of osmotic constituents of *Eriocheir* muscle is known in great detail makes a comparison useful with other marine forms, also approximately in equilibrium with seawater, i.e., with a solution containing approximately 1000 milliosmoles per liter of water, such as *Carcinus maenas* (Shaw, 1958) and *Nephrops norvegicus* (Robertson, 1961), and of the axoplasm of the giant nervous fibers of the squid *Loligo pealei* (Deffner and Hafter, 1960). This has been done in a recent review published by Florkin and Schoffeniels (1965). If we remain in the domain of the muscles of Crustacea, the inorganic composition appears as relatively more important in *Eriocheir* than in *Carcinus* and in *Nephrops*. With respect to the organic constituents, the sum of free amino acids is smaller in *Eriocheir* than in *Carcinus* and in *Nephrops*, while the concentrations of trimethylaminoxide is of the same order in *Eriocheir*, in *Nephrops*, and in *Carcinus*.

We are also well informed with respect to the nonprotein nitrogenous constituents of the abdominal muscle of the lobster *Homarus vulgaris* (Camien *et al.*, 1951; Kermack *et al.*, 1955). Taurine, trimethylaminoxide, and betaine are less concentrated in *Eriocheir* in seawater, than in the lobster. The pattern is nevertheless very similar in both cases, the sum of free amino acids being a little higher in the lobster, as well as the concentration of proline. The pattern in Crustacean muscles is quite different from that observed in the giant nerve fibers of *Loligo*, glycine, proline, glutamic acid, and alanine, which form the main constituents of the organic osmotic effectors in Crustacea, since they are of little weight in the *Loligo* fibers where aspartic acid predominates and where isethionic acid and glycerol are significant constituents.

In the case of *Eriocheir sinensis*, an anisosmotic extracellular regulation takes place in fresh water and in brackish water. This regulation limits the change of blood osmotic pressure. This change, therefore, is smaller than it would be if the equilibrium was established between blood and environment at all concentrations of the latter.

When the external medium is diluted, there is a certain degree of blood dilution. This produces, as a consequence, a change in the degree of hydration of the cell contents. This change is nevertheless kept to a small degree of magnitude by the isosmotic intracellular regulation which exerts an important, if not complete, adjustment of the cell content to the new blood concentrations. The coexistence of the two mechanisms (anisosmotic extracellular regulation and isosmotic intracellular regulation) can also be observed in euryhaline species such as *Carcinus maenas* (Duchâteau and Florkin, 1956; Shaw, 1958; Duchâteau *et al.*, 1959), *Leander serratus* and *Leander squilla* (Jeuniaux *et al.*, 1961a), or *Nereis diversicolor* (Jeuniaux *et al.*, 1961b), as well as in euryhaline teleosts (Lange and Fugelli, 1965). It is also present in terrestrial or semiterrestrial species such as *Cardisoma armatum* (De Leersnijder and Hoestlandt 1963-1964), *Pachygrapsus crassipes* (Jones, 1941; Gross, 1955, 1957), *Birgus latro* (Gross, 1955, 1957), *Coenobita perlatus* (Gross and Holland, 1960), and *Gecarcinus lateralis* (Gross, 1963).

The euryhalinity of *Eriocheir sinensis* and of *Carcinus maenas* are well known and their mechanisms of anisosmotic extracellular regulation have been extensively studied. *Leander serratus* and *Leander squilla* have, in seawater, a blood hypotonic with respect to seawater. When transferred to brackish water, the difference between blood and medium diminishes and the blood is isotonic with the external medium when the Δ of the latter is lowered to 1.6°C. Below this value the blood is maintained hypertonic with respect to the brackish water, by the action of the anisosmotic extracellular regulation (Panikkar, 1940; Verwey, 1957).

Nereis diversicolor is the only marine invertebrate known, outside Crustacea, to show an anisosmotic extracellular regulation when it lives in brackish water where it can stand a five times diluted seawater (Schlieper, 1929).

On the other hand, the anisosmotic extracellular regulation is lacking in some evidently euryhaline invertebrates of the sea. Among Annelids, two species are often opposed with respect to their power of osmoregulation. *Nereis diversicolor* is very euryhaline, while the degree of euryhalinity of *Perinereis cultrifera*, which is found in certain estuaries (Hesse *et al.*, 1937), is more limited. Its limit of experimental tolerance corresponds to the concentration of twice diluted seawater, in which the change of the degree of tissue hydration remains small. This is not due to any anisosmotic extracellular regulation (Wells and Ledingham, 1940). When *Perinereis cultrifera* is transferred from seawater to twice diluted seawater, the change of Δ, in the internal medium as well as in the external medium, corresponds to about 1°C. But the isosmotic intracellular regulation osmotically adjusts the intracellular contents to the new situation and no excessive change of hydration takes place in the cells (Jeuniaux *et al.*, 1961b).

It is only on the intracellular isosmotic regulation that the euryhalinity of *Arenicola marina* (Duchâteau-Bosson *et al.*, 1961) of *Mytilus edulis* (Potts, 1958; Bricteux-Grégoire *et al.*, 1964a), of *Ostrea edulis* (Bricteux-Grégoire *et al.*, 1964b), of *Gryphea angulata* (Bricteux-Grégoire *et al.*, 1964c), of *Asterias rubens* (Jeuniaux *et al.*, 1962), of *Golfingia gouldii* (Virkar, 1966), and *Strongylocentrotus droebachiensis* (Lange, 1964) depend. *Arenicola marina* shows no anisosmotic blood regulation (Schlieper, 1929; Beadle, 1937; Zenkewitch, 1938), though its distribution extends into brackish waters. *Mytilus edulis* shows no anisosmotic extracellular regulation (Potts, 1958). As observed by Binyon (1961), there is no such regulation either in *Asterias rubens*. Nevertheless the populations of this species extend, in the Baltic Sea, as far as Rugen Island, in the vicinity of which the salinity is of 8/1000 (Brattsbröm, 1941; Segersträle, 1949; Schlieper, 1957). Binyon (1961) has shown that in the North Sea the populations of *Asterias rubens* can stand, for a limited time, a salinity as low as 23/1000.

In all cases of euryhaline marine invertebrates so far studied, the intracellular isosmotic regulation is found, while the anisosmotic extracellular regulation, relieving the intracellular mechanism of a part of its duty, is not always present. We must therefore consider the intracellular isosmotic regulation as the more primitive mechanism, to which, in the species with a more extended euryhalinity, the anisosmotic regulation adds a new range of possibilities. In all the cases examined, the reversible variation of the free amino acid concentration, which can only be explained to a limited extent, by the change of hydration, contributes for a notable part to the change of intracellular osmoconcentration.

What are the amino acids taking part in the regulation? It can be concluded that most of the effect is due to low molecular weight amino acids: alanine, glycine, aspartic and glutamic acids, and proline. They are not however the only effectors of the intracellular isosmotic regulation. The inorganic constituents play a role which varies from one case to another. Taurine is an important effector in the isosmotic regulation of *Asterias rubens* (Jeuniaux *et al.*, 1962), of *Mytilus edulis,* and of *Gryphea angulata* (Bricteux-Grégoire *et al.*, 1964a,c), while it does not appear to take part in the process as it is accomplished in *Eriocheir sinensis, Leander serratus, Leander squilla,* or *Ostrea edulis* (Bricteux-Grégoire *et al.*, 1964b). Trimethylaminoxide is one of the effectors of the regulation in *Carcinus maenas* (Shaw, 1958), in *Eriocheir sinensis* (Bricteux-Grégoire *et al.*, 1962), in *Pleuronectes flesus,* and in *Gasterosteus aculeatus* (Lange and Fugelli, 1965).

Euryhalinity is also sometimes observed in freshwater invertebrates. In spite of the fact that the crayfish is a typical freshwater form, there are a number of records of crayfish in brackish water (Smith, 1959). In an experimental study by Herrmann (1931), it was observed that the crayfish *Astacus astacus* can support an increase in concentration of the water up to $\Delta = 1.07°C$, where the animals live for a week, whereas they rapidly die if placed in more concentrated media. Values for the concentrations of free amino acids in blood plasma and muscles of *Astacus astacus* have been recorded by Camien *et al.* (1951) and by Duchâteau and Florkin (1954, 1961). These determinations have been performed on dialyzates of inactivated homogenized muscles, before and after hydrolysis. As the hydrolysis changes the concentration of aspartic and glutamic acid only, the data represent the values for the concentration of free amino acids, except in the case of glutamic and aspartic acids, where they correspond to the sum of the free amino acid and of its amide. Similar results have been obtained for another crayfish, *Astacus (Austropotamobius) pallipes* Lereboullet (= *Saxatilis* Bell) by Cowey (1961). When *Astacus astacus* lives in fresh water ($\Delta = 0.02°C$), the Δ of the blood is maintained at a much higher value (0.8°C) than that of the external medium (Fredericq, 1898). In all the media in which the crayfish survive (i.e., up to a Δ of about 1.07°C in the external medium), the Δ of the blood is higher than the Δ of the external medium and there is no sensible change in weight (Herrmann, 1931). In fresh water, the great difference of concentration between the external medium and the cell fluid, which is itself in equilibrium with the blood, is partly compensated by the mechanism of anisosmotic extracellular regulation maintaining the Δ of the blood at a higher value. This value is still smaller than that found in the lobster, whose blood is osmotically in equilibrium with seawater. The smaller intracellular Δ for the crayfish is partly due to the smaller concentration of free amino acids in the cells. Anisosmotic extracellular regulation is not a factor of eury-halinity when the crayfish is transferred to more concentrated media. On the contrary, it maintains a concentration in the blood higher than in the medium, which increases the osmotic discrep-ancy between cells and blood. In the crayfish, an isosmotic intra-cellular regulation compensates for the increase in blood con-centration of the blood and avoids extended changes of cell water content (Tables 6-4 and 6-5).

Table 6-4 shows the changes in sodium and potassium content in twice diluted seawater, measured in the blood and in abdomi-nal muscles, after three weeks in the concentrated medium. The degree of hydration of the muscles shows only a very slight mod-ification after two weeks, 83.3 and 83.8% water in fresh and brackish water respectively, in spite of the fact that the Δ of the water has been changed from $0.02°$ to $0.95°C$ and the Δ of the in-ternal medium from $0.8°$ to $1.04°C$. There is an increase of con-centration of the total of the free amino acids determined. The mean increase amounts to 82.9 moles of amino acids per kg of water ($252.5 - 169.6$), which corresponds to a chance of Δ of $0.16°$ C. As the increase in the blood concentration, in these circum-stances, corresponds to $0.24°C$ ($1.04° - 0.8°C$), the increase in the concentration of the 15 amino acids determined compensates osmotically for more than half the increase of blood salts concen-tration. The variation of the concentration of the inorganic con-stituents of the muscle fibers contributes also to the compensa-tory mechanism in the crayfish. Among the amino acids, total glutamic acid + glutamine, proline, and alanine are responsible for the main part of the increase of intracellular osmotic concen-tration in the concentrated medium, due to the free amino acids.

Summarizing, it seems well demonstrated that the primitive function of the active transport of cations is that of cell volume regulation. Together with the control of the intracellular concen-tration of some products of nitrogen metabolism, it provides the cell with mechanisms enabling it to keep the intracellular water

TABLE 6-4
Astacus astacus: SERUM OR ABDOMINAL MUSCLE[a]

Inorganic constituents	Serum (mOsm/liter)		Muscles (mOsm/liter water)	
	Controls in fresh water	50/100 Sea-water (15 days)	Controls in fresh water	50/100 Sea-water (15 days)
Potassium	5.4	5.2	103.0	115.0
Sodium	172.0	212.0	52.0	70.0

[a]From Schoffeniels, unpublished results.

TABLE 6-5
Astacus astacus: ABDOMINAL MUSCLE[a]

Amino acid	Mean content (mOsm/1000 gm fresh tissue)		Mean content (mOsm/liter of water)	
	Controls in fresh water 10−11°C October 5, 1955	50 per 100 seawater 10−11°C September 20, 1955− October 5, 1955	Controls in fresh water idem	50 per 100 seawater idem
Ala	13.3	21.1	16.1	26.0
Arg	37.0	44.2	44.8	54.5
Asp	8.0	9.5	9.7	11.7
Glu	34.4	50.5	41.7	62.3
Gly	20.9	24.3	25.3	30.0
His	0.6	1.6	0.7	2.0
Ileu	1.7	3.4	2.1	4.2
Leu	2.0	4.0	2.4	4.9
Lys	5.6	7.0	6.8	8.6
Met	1.7	1.8	2.1	2.2
Phe	0.9	0.6	1.1	0.7
Pro	7.8	25.7	9.5	31.7
Thr	2.7	5.6	3.7	6.9
Tyr	1.2	1.0	1.5	1.2
Val	2.2	4.6	2.7	5.7
Total	140.0	204.9	170.2	252.6

[a]From Duchâteau-Bosson and Florkin, 1961.

activity nearly equal to that of the extracellular medium. As a consequence we may consider that the cell osmotic pressure is always very near to that of the extracellular medium.

In the particular case of the euryhaline species, this seems also to hold true since the total concentration of small intracellular solutes varies parallelly to that found in the extracellular fluid. Thus the cell behaves as if its osmotic pressure does not exceed that of the extracellular solution. Therefore the concept of isosmoticity is of real operational value as such and provided that it is clearly understood that more has still to be known about the activity of water inside the cell before asserting that it is exactly equal to that of the extracellular fluid.

The way the cell controls the mechanisms enabling it to keep its volume constant is far from being completely understood. Experiments done in order to elucidate this problem are presented in the next chapter.

The interesting suggestion of Kleinzeller as to the existence of a contractile mechanism in cell volume regulation requires further experimental investigations in order to define the cell categories and the exact conditions in which such a mechanism could be operative.

References

Beadle, L. C. (1937). *J. Exptl. Biol.* **14**, 56.

Binyon, J. (1961). *J. Marine Biol. Assoc. U.K.* **41**, 161.

Brattsbröm, H. (1941). *Undersöknigar over Oresund*, Lund. **27**.

Bricteux-Grégoire, S., Duchâteau-Bosson, Gh., Jeuniaux, Ch., and Florkin, M. (1962). *Arch. Intern. Physiol. Biochim.* **70**, 273.

Bricteux-Grégoire, S., Duchâteau-Bosson, Gh., Jeuniaux, Ch., and Florkin, M. (1964a). *Arch. Intern. Physiol. Biochim.* **72**, 116.

Bricteux-Grégoire, S., Duchâteau-Bosson, Gh., Jeuniaux, Ch., and Florkin, M. (1964b). *Arch. Intern. Physiol. Biochim.* **72**, 267.

Bricteux-Grégoire, S., Duchâteau-Bosson, Gh., Jeuniaux, Ch., and Florkin, M. (1964c). *Arch. Intern. Physiol. Biochim.* **72**, 835.

Camien, M.N., Sarlet, H., Duchâteau, Gh., and Florkin, M. (1951). *J. Biol. Chem.* **193**, 881.

Cole, K.S. (1932). *J. Cellular Comp. Physiol.* **1**, 1.

Cowey, C.B. (1961). *Comp. Biochem. Physiol.* **2**, 173.

Deffner, G. G. J., and Hafter, R. E. (1960). *Biochim. Biophys. Acta* **42**, 200.

De Leersnijder, M. (1966). "Influence de quelques facteurs externes et internes sur le milieu intérieur, la mue et le développement ovarien d'*Eriocheir sinensis* Milne Edwards." Doctoral thesis, Faculté des Sciences de Lille, France.

De Leersnijder, M. (1967a). *Cahiers Biol. Marine,* **8**, 195.

De Leersnijder, M. (1967b). *Cahiers Biol. Marine,* **8**, 295.

De Leersnijder, M., and Hoestlandt, H. (1963–1964). *Mem. Soc. Sci. Nat. Math. Cherbourg,* **51** (6e sér.), 43.

Deyrup, I. (1953). *J. Gen. Physiol.* **36**, 739.

Duchâteau, Gh., and Florkin, M. (1954). *Arch. Intern. Physiol.* **62**, 487.

Duchâteau, Gh., and Florkin, M. (1955). *Arch. Intern. Physiol. Biochim.* **63**, 249.

Duchâteau, Gh., and Florkin, M. (1956). *J. Physiol. Paris* **48**, 520.

Duchâteau-Bosson, Gh., and Florkin, M. (1961). *Comp. Biochem. Physiol.* **3**, 245.

Duchâteau-Bosson, Gh., and Florkin, M. (1962a). *Arch. Intern. Physiol. Biochim.* **70**, 345.

Duchâteau-Bosson, Gh., and Florkin (1962b). *Arch. Intern. Physiol. Biochim.* **70**, 393.

Duchâteau-Bosson, Florkin, M., and Jeuniaux, Ch. (1959). *Arch. Intern. Physiol. Biochim.* **67**, 489.

Duchâteau-Bosson, Gh., Jeuniaux, Ch., and Florkin, M. (1961). *Arch. Intern. Physiol. Biochim.* **69**, 30.

Eisenberg-Hamburg, E. (1925). *Arch. Biol.* **35**, 441.

Florkin, M. (1956). "Vergleichend. biochemischen Fragen. 6. Colloquium der Gesellschaft für physiologische Chemie" pp. 62–99. Springer, Berlin.

Florkin, M., and Schoffeniels, E. (1965). *In* "Studies in Comparative Biochemistry" (K. A. Munday, ed.) pp. 6–40. Pergamon, Oxford.

Franck, G., Cornette, M., and Schoffeniels, E. (1968). *J. Neurochem.* **15**, 843.

Fredericq, L. (1898). *Bull. Acad. Roy. Belg.* **35**, 831.

Fugelli, K. (1967). *Comp. Biochem. Physiol.* **22**, 253.

Gross, W. J. (1955). *Am. Naturalist* **89**, 205.

Gross, W. J. (1957). *Biol. Bull.* **112**, 43.

Gross, W. J. (1963). *Physiol. Zool.* **36**, 312.

Gross, W. J., and Holland, P. V. (1960). *Physiol. Zool.* **33**, 21.

Hesse, R., Allee, W. C., and Schmidt, K. P. (1937). "Zoological Animal Geography." Wiley, New York.

Herrmann, F. (1931). *Z. vergleich. Physiol.* **14**, 479.

Jeuniaux, Ch., Bricteux-Grégoire, S., and Florkin, M. (1961a). *Cahiers Biol. Marine* **2**, 373.

Jeuniaux, Ch., Duchâteau-Bosson, Gh., and Florkin, M. (1961b). *J. Biochem.* **49**, 527.

Jeuniaux, Ch., Bricteux-Grégoire, S., and Florkin, M. (1962). *Cahiers Biol. Marine* **3**, 107.

Jones, L. L. (1941). *J. Cellular Comp. Physiol.* **18**, 79.

Kermack, W. O., Lees, H., and Wood, J. D. (1955). *Biochem. J.* **60**, 424.

Kleinzeller, A., and Knotkova, A. (1964a). *J. Physiol. London* **175**, 172.

Kleinzeller, A., and Knotkova, A. (1964b). *Physiol. Bohemoslovenica* **13**, 317.

Lange, R. (1964). *Comp. Biochem. Physiol.* **13**, 205.

Lange, R., and Fugelli, K. (1965). *Comp. Biochem. Physiol.* **15**, 283.

Leaf, A. (1956). *Biochem. J.* **62**, 241.

Lehninger, A. L. (1962). *Physiol. Rev.* **42**, 467.

Mudge, G. H. (1951). *Am. J. Physiol.* **165**, 113.

Ohnishi, T. (1962). *J. Biochem. Tokyo* **52**, 307.

Opie, E. L. (1949). *J. Exptl. Med.* **89**, 185.

Panikkar, N. K. (1940). *Nature* **146**, 366.

Potts, W. T. W. (1958). *J. Exptl. Biol.* **35**, 749.

Rand, R. P., and Burton, A. C. (1964). *Biophys. J.* **4**, 491.

Raze, C., and Schoffeniels, E. (1965). *Bull. Acad. Roy. Belg., Classe Sci.* **51**, 1057.

Robertson, J. D. (1961). *J. Exptl. Biol.* **38**, 707.

Robinson, J. R. (1960). *Physiol. Rev.* **40**, 112.

Rothstein, A. (1959). *Bacteriol. Rev.* **23**, 175.

Schlieper, C. (1929). *Z. vergleich Physiol.* **9**, 478.

Schlieper, C. (1957). *Année Biol.* **33**, 117.

Schoffeniels, E. (1961). *In* "Comparative Study of Membrane Permeability. Biological Structure and Function" (T. W. Goodwin and O. Lindberg, eds.) Vol. II, pp. 621—631 Academic Press, London.

Schoffeniels, E. (1964). *In* "Comparative Biochemistry" (M. Florkin and H. S. Mason, eds.) Vol. 7, pp. 137—202. Academic Press, New York.

Schoffeniels, E. (1967). "Cellular aspects of membrane permeability." Pergamon, Oxford.

Schoffeniels, E., and Tercafs, R. (1965—1966). *Ann. Soc. Roy. Zool. Belg.* **96**, 23.

Scholles, W. (1933). *Z. vergleich. Physiol.* **19**, 522.

Segerstråle, S. (1949). *Oikos* **1**, 127.

Shaw, J. (1958). *J. Exptl. Biol.* **35**, 920.

Smith, R. I. (1959) *In* "Marine Biology (Proc. 20th Ann. Biol. Coll. of Oregon State College, Corvallis, Oregon)" pp. 59—69.

Tosteson, D. C. (1964). *In* "The Cellular Functions of Membrane Transport" (J. F. Hoffman, ed.) pp. 3-22. Prentice-Hall, New York.

Verwey, J. (1957). *Année Biol.* **33**, 129.

Virkar, R. A. (1966). *Comp. Biochem. Physiol.* **18**, 617.

Webber, H. H., and Dehnel, P. A. (1968). *Comp. Biochem. Physiol.* **25**, 49.

Wells, G. P., and Ledingham, I. C. (1940). *J. Exptl. Biol.* **17**, 337.

Whittam, R. (1956). *J. Physiol. London* **131**, 542.

Wins, P., and Schoffeniels, E. (1965). *Arch. Intern. Physiol. Biochim.* **73**, 160.

Wins, P., and Schoffeniels, E. (1966a). *Biochim. Biophys. Acta* **120**, 341.

Wins, P., and Schoffeniels, E. (1966b). *Arch. Intern. Physiol. Biochim.* **74**, 812.

Wins, P., and Schoffeniels, E. (1967). *Biochim. Biophys. Acta,* **135**, 831.

Wohlfarth-Bottermann, K. E. (1963). *Naturwissenschaften* **50**, 237.

Zenkewitch, L. (1938). *Zool. Zh.* **17**, 845 (cited by Wells and Ledingham, 1940).

CHAPTER VII

Anisosmotic Extracellular and Isosmotic Intracellular Regulations as Examples of Physiological Radiations of Biochemical Systems

As mentioned earlier in the text, when considering the adaptation of a species to a given environment, the resulting situation very often depends on the properties of systems that range from the macroscopic down to the molecular scale. The special way moths are adapted to escape from cobwebs — eluding capture by losing their scales to the viscid threads (Eisner *et al.*, 1964) — if it indicates that cuticle with hairs or scales are those from which the spider thread detaches most readily, has however little molecular content. This adaptation cannot be traced down to special properties of a molecular species but rather depends on the morphological structure of the integument, a property that is undoubtedly polygenic. The possibility of an aquatic animal to succeed in media of various salinities is also dependent on the existence of different systems, some mechanical others molecular, working synergistically, which defines the degree of euryhalinity of the species under consideration. In the case of some mollusks, such as *Littorina littorea* and *Purpura lapillus*, the perivisceral fluid acts as a buffer between the blood and the external medium when placed in a dilute medium, the operculum closes the opening of the shell, which offers a temporary means of preventing the effects of too rapid a change in the composition of the perivisceral fluid (Hoyaux, 1967). However, life in a dilute

112

medium is only possible if the animal possesses the molecular adaptations that enable it to keep not only the concentration and the *composition* of its blood in the adequate state, but also its cell volume constant.

Euryhalinity, or adaptation to a given environment, cannot be explained by a single molecular property, but rather results from the association of mechanisms whose hierarchy has to be evaluated to understand fully the phenomenon. This explains why some species, such as *Maia squinado*, can be adapted in the laboratory to media differing considerably from their natural environment. It is important to consider what could be called ecological euryhalinity as opposed to an experimental or physiological one.

The existence of the freshwater forms and of some marine fishes like the teleosts is accounted for by an anisosmotic extracellular regulation. Anisosmotic regulation also contributes to the euryhalinity of some forms. In the anisosmotic regulation of the body fluids, we are concerned with a steady state between the external medium and the internal medium. In the animals maintaining a hyperosmotic state, the steady state in the body fluids is the resultant of discharge phenomena and of charge phenomena, the discharge being the result of a loss of ions by diffusion through the body and by removal of urine, and the charge being ensured by absorption. This absorption sometimes takes place at the digestive frontier, but in many cases, specialized mechanisms have been developed for active absorption of ions from dilute solution. Even in starvation, amphibians can maintain the steady state of their body fluids. When kept in distilled water, a frog will lose salts through the skin and through urine but it will take up sodium chloride to restore the initial blood concentration if transferred to a solution of sodium chloride as diluted as tapwater. The skin, in which the absorption mechanism has been located, can absorb for several hours when isolated. Sodium is actively transported through the skin; the process generates a resting potential which is in turn responsible for the passive displacement of chloride ions (Ussing and Zerahn, 1951).

The Chinese crab *Eriocheir sinensis* has a very great power of ion absorption while living in fresh water. The salt loss takes place through the isosmotic urine and the body wall, notably the anterior gills. The absorption mechanism is located in the posterior gills. As shown by Koch (1954), the gill will continue to absorb ions *in vitro* some time after being separated from the body. It is even possible to perfuse an isolated gill and measure simultaneously efflux and influx of an ionic species. Knowing the electrical potential difference and the composition of the perfusate as well as that of the outside medium, it is possible to define the nature of the force responsible for the displacement of the ion under consideration. This was done in our laboratory by Dr. E. King for three species of crab. Table 7-1 shows the results of representative experiments dealing with the measurement of the sodium flux. When considering the same species, flux values varied from preparation to preparation, the lowest values being associated with considerable amount of blood clot. This reduced the area available for exchange, thus explaining the results obtained. No conclusions should be reached with regard to species differences in flux, with the exception that the efflux values from the posterior gills of *Eriocheir* were an order of magnitude lower than those obtained in the other species. Changes in the flux values within a preparation are considered significant and flux ratios may of course be compared. The most significant conclusions that may be drawn from the consideration of Table 7-1 are the following. In *Maia*, the sodium influx varies as a function of the concentration of the external medium. The efflux decreases somewhat when the external sodium is lowered, the greatest change occurring between 450 and 340 μmoles/ml. There is no measurable transepithelial electrical potential in isosmotic medium. In dilute media, the potential difference is only one millivolt, as the blood is negative to the medium.

When the flux ratio is calculated according to Ussing (1949) and compared to the value found experimentally, there is good agreement in isosmotic saline and in the two most dilute media. This firmly supports the conclusion that the movement of sodium is a passive phenomenon. When the gills are placed in a medium corresponding to 75% seawater, the observed ratio is 0.9 while the calculated value is 0.75. This suggests that there may be a limited regulation of internal sodium as long as the medium is only slightly diluted, a reduction in efflux being primarily responsible for the observed result. The fact that *Maia* is able to live in 50% seawater for some time is easily explained by the above result and the observation that it possesses the isosmotic intracellular regulation (Schoffeniels, unpublished).

In *Carcinus*, Table 7-1 shows that both influx and efflux of sodium decreased as the external medium is diluted down to 75%. The flux ratios are approximately 1.0 at all the dilutions tested. A transepithelial potential develops when the gills are placed in dilute media and this potential agrees closely with that calculated from the Nernst equation for sodium diffusion. Since the sign of the potential indicates the blood is negative with respect to the medium, a passive influx of sodium would be expected. However since the observed flux ratios are always higher than that calculated according to Ussing (1949), it may be concluded that part of the influx must be an active phenomena. It is interesting to note that despite the fact that the sodium concentration of the perfusate is kept constant, the efflux values go down as the outside concentration decreases. This may point to the existence of an exchange diffusion for sodium in the gills of *Carcinus*.

In *Eriocheir* Koch (1954) has reported that only the posterior three pairs of gills show active transport of sodium; the anterior pairs were found to be less effective. The results of Dr. E. King confirm and extend these findings. In the posterior gills, the efflux of sodium decreases as the medium is diluted. It is always very low at all the dilutions tested. A potential difference of 15 to 30 millivolts is observed. It is independent of the concentration of the external salt solution. The observed flux ratio is always greater than the calculated ratio and frequently exceeds it by tenfold. For example, note the value for posterior gill "b" in 2% seawater (Table 7-1), a result that undoubtedly indicates that salt absorption is an active phenomena.

In the anterior gills of *Eriocheir*, while influx values are comparable to those of the posterior gills, the efflux values are of a larger order of magnitude. In dilute media, a potential difference, which agrees closely with that calculated for a sodium diffusion potential, develops. The flux ratio values indicate that a mechanism for active transport of sodium may be operative. It remains however difficult to explain the function of the anterior gills in the sodium balance of *Eriocheir*. It is possible that the

TABLE 7-1

SODIUM FLUX IN CRAB GILLS[a]

			Conditions							

Na_{out}/Na_{in}		1.0						.76		
Medium (μmoles/ml)		450						340		
Perfusate (μmoles/ml)		450						450		

				Ratio					Ratio	
Crab species	In	Out	PD	Experi-mental	Calcu-lated	In	Out	PD	Experi-mental	Calcu-lated
Maja	4.2	4.2	0	1.0	1.0	3.2	3.5	0	.91	.76
Carcinus	.86	.80	+1.0	1.1	.95	–	–	–	–	–
	1.16	1.0	+1.0	1.2	.95	1.13	1.18	−3.0	.96	.83

			Conditions							

Na_{out}/Na_{in}		1.0						.75		
Medium (μmoles/ml)		270						202		
Perfusate (μmoles/ml)		270						270		

				Ratio					Ratio	
Crab species	In	Out	PD	Experi-mental	Calcu-lated	In	Out	PD	Experi-mental	Calcu-lated
Eriocheir[b]										
Anterior gill	a 1.15	1.1	+3.0	1.0	.87	–	–	–	–	–
	b .38	.26	+3.0	1.4	.87	–	–	–	–	–
Posterior gill	a .14	.06	−15.0	2.3	1.8	–	–	–	–	–
	b .68	.02	−29.0	28.0	2.9	–	–	–	–	–

[a]After E. King, unpublished data. Representation experiments are given for each of the crabs. Perfusate was a balanced salt solution (Cole, 1941, in Lockwood, 1961). Medium was composed of the balanced salt solution diluted with distilled water as indicated. Flux values are given as micromoles per minute per gill. In = Appearance of $^{22}NaCl$ in the perfusate; Out = appearance of $^{24}NaCl$ in the medium; PD = potential difference measured by inserting capillary elec-

Conditions														
.4					.2					.02				
180					90					9				
450					450					450				
			Ratio					Ratio					Ratio	
In	Out	PD	Experimental	Calculated	In	Out	PD	Experimental	Calculated	In	Out	PD	Experimental	Calculated
1.3	3.1	− 0.5	.43	.41	.8	2.9	−1.0	.27	.21	−	−	−	−	−
.7	.71	−14.0	.98	.70	.55	.56	−24.0	.98	.5	−	−	−	−	−
										−	−	−	−	−

Conditions														
.5					.2					.02				
135					54					5				
270					270					270				
			Ratio					Ratio					Ratio	
In	Out	PD	Experimental	Calculated	In	Out	PD	Experimental	Calculated	In	Out	PD	Experimental	Calculated
0.82	1.1	− 5.0	.78	.60	.53	.78	−27.0	.65	.58	−	−	−	−	−
−	−	−	−	−	.26	.29	−27.0	.88	.58	.15	.37	−58.0	.39	.2
0.12	.09	−15.0	1.3	.90	.08	.1	−16.0	.80	.37	−	−	−	−	−
−	−	−	−	−	.35	.04	−25.0	8.8	.50	.13	.06	−25.0	2.1	.05

trodes connected to calomel half-cells into the perfusate and the medium. Electrodes were balanced in the medium. Sign is given as blood relative to medium. Calculated ratio (in/out) $= (Na_{out}/Na_{in}) \, e^{zF\Delta E/RT}$.

[b]Anterior and posterior gills "a" are from the same animal, and determinations were made simultaneously. The same is true for gills "b." Measurements were taken at 10−20 minute intervals, and values represent the mean of 2−3 readings.

blood is differentially directed toward the anterior or the posterior gills according to the external conditions of salinities, which points to the possibility of an adaptation of the circulatory system in the euryhaline abilities of this species.

It should also be mentioned that in the experiments reported in Table 7-1 the sodium concentration of the perfusate has never been changed over the whole range of concentrations tested for the external medium. It is thus not surprising that the efflux values are little affected (e.g., in *Maia* or in the anterior gills of *Eriocheir*). This situation is not encountered normally, since the blood composition is modified when the animal goes into media of different concentrations. The passive component of the efflux should thus decrease, with the consequence that a still higher flux ratio should obtain. At any rate the results show that in the case of *Carcinus* and *Eriocheir* an active transport of sodium located in the gills explains how these species may keep their ionic balance in dilute media.

These active absorption phenomena, which are at work in the gills of *Eriocheir* or *Carcinus* for instance, or in frog skin, are physiological radiations[1] of the biochemical system which ensures, inside most cells, the presence of high potassium and low sodium concentrations.

It appears that in most of the cases so far studied a system located in the membrane and requiring ATP is at work and helps to keep the intracellular sodium concentration low. If an epithelial cell acquires, as in the frog skin, the renal tubule, or the gills of crustacea, the capacity to transport sodium from one side of the epithelium to the other, we may consider this as being a molecular evolution in the nature of a physiological radiation of the system present in all cells — this radiation consists of a spatial distribution of permeability characteristics. This interpretation, proposed by Schoffeniels (1959), applies to frog skin (Koefoed-Johnsen and Ussing, 1958): the cellular membranes facing outward lose their capacity for transporting sodium actively but al-

[1]Homologous biochemical molecules or biochemical systems may show, whether or not they have undergone change in descent or a change of activity, or even if the activity remains unchanged at the molecular level, take part in systems subserving other functions at higher organization levels. This defines the concept of physiological radiation (Florkin, 1959).

low sodium to enter the cell by diffusion. As a result there is a net transfer of sodium across the epithelium in the inward direction.

This scheme corresponds also to the asymmetry in the biochemical properties of other epithelia such as those found in the digestive tract, the respiratory system etc. (Schoffeniels, 1967a). It also explains the origin of the action potential in conductive cells and how a powerful electrical discharge may be produced by the gymnote, the torpedo, and other electric fishes (Schoffeniels, 1959, 1964a, 1967a; Whittam and Guinnebault, 1960; Bartels *et al.*, 1960; Rosenberg and Higman, 1960; Whittam, 1961; Bartels, 1962). The mechanism of anisosmotic extracellular regulation appears therefore as one of the physiological radiations of the system controlling the permeability of the cell membrane. This system is uniformly distributed at the surface of the cell in the primitive stages of its functioning and its physiological radiation which is observed in the absorption through cell membranes, in the conduction along axons, in electrical discharges in fishes, etc. depends on changes of distribution and organization of the permeability characteristics of the cell boundary (Schoffeniels, 1959, 1961, 1967a)

When we consider the aspects of isosmotic intracellular regulation as they take place in a crustacean, such as *Eriocheir sinensis*, we see that in this regulation a change in distribution takes place; this is another physiological radiation of the system controlling membrane permeability.

As we have shown in the preceding chapter, in tissues of crustacea, the greatest part of the contribution to the intracellular component, as well as the contribution to the intracellular adjustment when dealing with an euryhaline species, is due to nonessential amino acids, such as alanine, glycine, glutamic acid, and proline. The question now arises as to the origin of the amino acids. There are at least two possibilities; the amino acids could be of extracellular origin and transported (actively or not) in the cell; or they could come from within the cell.

On the basis of experiments performed on isolated nerves of *Eriocheir sinensis*, it has been concluded (Schoffeniels, 1960): (1) that the intracellular isosmotic regulation is not dependent on hormonal mechanism; (2) that the amino acids contributing to the total osmotic pressure are of intracellular origin;

TABLE 7-2

COMPARISON OF FREE AND TOTAL ALANINE AND PROLINE IN THE MUSCLE OF *Eriocheir sinensis* ADAPTED TO FRESH WATER AND SEAWATER[a]

Crab number	Free alanine (mg/100 gm dry muscle)			Total alanine (mg/100 gm dry muscle)		
	After 6 days			After 6 days		
	Fresh water	in seawater	Variation	Fresh water	in seawater	Variation
1	1764	2899	+1135	5582	6777	+1195
2	–	–	–	–	–	–
3	1794	3183	+1389	5633	6594	+ 961

[a] After Florkin *et al.*, 1964.

and (3) that the osmotic pressure per se is not responsible for the increase in amino acid concentration; the presence of sodium or potassium is necessary. This is why it has been suggested that the ionic content of the cell could be responsible for the regulation of the amino acid metabolism (Schoffeniels, 1959).

But these results leave another question unanswered. The regulation of the intracellular pool of amino acids could indeed be mainly dependent on the turnover rate of some proteins or result from a balance between synthesis and breakdown of amino acids. By measuring the variation in free amino acids before and after hydrolysis of proteins, an indication may be obtained as to the possible participation of protein hydrolysis in the cell adaptation to an increase in the osmotic pressure of the blood.

Table 7-2 shows that when *Eriocheir* is adapted to seawater, the increase in free alanine is accompanied by a parallel increase in the total alanine. The results also show that while the free proline has markedly increased in the muscle, the amount of total proline has increased proportionally, which indicates that the proline obtained from the hydrolysis of proteins did not vary significantly.

These results are in agreement with the conception of a net synthesis of alanine and proline during the adaptation to seawater rather than a liberation from some tissular reserve.

On the other hand, in collaboration with Dr. G. Hamoir, we

Free proline (mg/100 gm dry muscle)			Total proline (mg/100 gm dry muscle)		
	After 6 days			After 6 days	
Fresh water	in seawater	Variation	Fresh water	in seawater	Variation
1966	2575	+ 609	4451	4614	+ 163
953	1755	+ 802	3783	4531	+ 748
718	1446	+ 728	3707	4569	+ 862

have examined the protein composition of *Eriocheir* muscle and have been unable to detect any change in the electrophoretic pattern of the proteins during the adaptation of *Eriocheir sinensis* to seawater. Osmotic adaptation is also paralleled by a modification of nitrogen excretion which increases when a euryhaline species is transferred to diluted medium and decreases during the adaptation to hypertonic medium. This has been demonstrated for *Carcinus maenas* (Needham, 1957) and *Eriocheir sinensis* (Fig. 7-1 and 7-2).

From the above experiments, it is likely that the regulation of the intracellular amino acid pool may depend on the acquisition of mechanisms controlling the relative speeds of the anabolism and catabolism of the amino acids concerned. It was therefore imperative to learn more about the relationship between carbohydrate metabolism and the metabolism of amino acids such as alanine, glycine, and glutamic acid, in the tissues of crustacea.

In a series of experiments, the incorporation of ^{14}C from glucose and pyruvate, in the free amino acids of the isolated nervous chain of the lobster *Homarus vulgaris* L. and of the crayfish *Astacus fluviatilis* L., was studied. The aim was to determine the pathways leading to the biosynthesis of amino acids in normal conditions as well as in experimental conditions known to affect the ionic distribution in the intracellular space. Two pharmacological agents, veratrine and cocaine, have been chosen—their action is generally interpreted in terms of an effect on the perme-

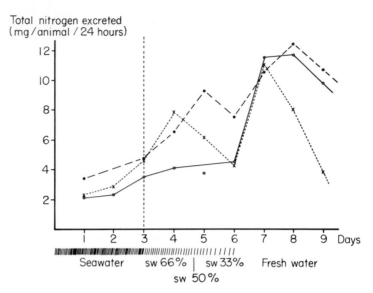

FIG. 7-1. Nitrogen excretion during the adaptation of *Eriocheir sinensis* to fresh water. Results obtained with three different individuals, expressed as total nitrogen excreted in milligrams per animal per 24 hour period as function of time. (Florkin *et al.*, 1964.)

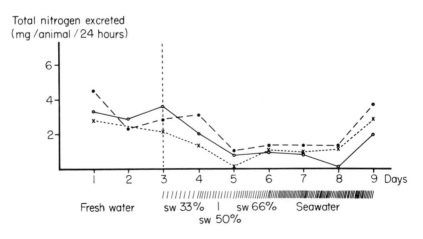

FIG. 7-2. Nitrogen excretion during adaptation of *Eriocheir sinensis* to seawater. Results obtained with three different individuals and expressed as in Fig. 7-1. (Florkin *et al.*, 1964.)

ability of the cell membrane. As a result there would be a shift in the ionic distribution between the cell and its surroundings.

If this is true as far as veratrine is concerned (Table 7-3) the results obtained with cocaine indicate little effect, if any, on the ionic composition of the ventral nerve chain. Cocaine however, like veratrine, has an effect on the amino acid metabolism (Gilles and Schoffeniels, 1964). Since it is also active on cell homogenates, contrary to veratrine, one has to postulate a direct action on one or several enzymatic systems. It is known, for example, that cocaine inhibits the introduction of the acetyl CoA into the Krebs cycle. This is in accordance with the fact that, in the experiments on the ventral nervous chain of the lobster a reduced synthesis of dicarboxylic acids was observed (Gilles and Schoffeniels, 1964).

For the study of the amino acid metabolism, the following general procedure is used. A ventral chain is incubated in an appropriate physiological saline for periods of 1 to 4 hours, with glucose-U-^{14}C, glucose-1-^{14}C, glucose-6-^{14}C, pyruvate-1-^{14}C or pyruvate-2-^{14}C; the functional state of the preparation was

TABLE 7-3

VENTRAL NERVE CHAIN OF *Homarus vulgaris* L. INFLUENCE OF 4 μM VERATRINE SULFATE AND 4mM COCAINE HYDROCHLORIDE ON THE CONCENTRATION OF Na AND K[a]

First half		Second half					
Control		Control		4 μM Veratrine sulfate		4 mM Cocaine	
Na	K	Na	K	Na	K	Na	K
4.896	3.727	4.987	4.206	—	—	—	—
5.987	4.003	5.850	3.893	—	—	—	—
5.512	4.013	—	—	6.617	2.686	—	—
5.951	3.903	—	—	6.935	2.536	—	—
6.106	4.274	—	—	—	—	6.875	3.955
5.003	3.292	—	—	—	—	5.363	3.627
5.238	3.473	—	—	—	—	5.238	3.015

[a]Results expressed in gm of the considered ion per kilogram of fresh weight. After Gilles and Schoffeniels, 1964.

checked at intervals during the experiment by testing the electrical activity. At the end of an incubation period, the amino acids are separated using an automatic analyzer supplemented with a scintillation counting unit placed on the effluent of the column.

The results obtained by Gilles and Schoffeniels (1966 and unpublished results) show that glycolysis and tricarboxylic acids cycle are operative in the ventral nerve chain of the lobster. The following enzymes have indeed been indentified: L-malate hydro-lyase (E.N. 4.2.1.2)[1] and L-malate: NAD oxidoreductase (E.N. 1.1.1.37). Moreover, the fact that in the presence of added L-malate a cellular extract of lobster or crayfish is able to reduce NADP indicates the presence of a malate dehydrogenase (L-malate:NADP oxidoreductase, E.N. 1.1.1.40) catalyzing reversibly the carboxylation of pyruvate. As the standard oxidoreduction potentials of the systems NADP(red)/NADP(ox) and malate/pyruvate $+CO_2$ are very close to each other (around -0.32 V) the reversibility of the reaction has a practical meaning. Oxaloacetate may thus be formed not only through the operation of the tricarboxylic acids cycle but also by carboxylation of pyruvate.

That this is the case may easily be demonstrated by studying the distribution of ^{14}C in amino acids obtained after incubation with glucose-^{14}C. The results show that the specific activities of glutamate and aspartate are very close to each other, a fact that may be interpreted as indicating either a compartmentation of 2-ketoglutarate or that a route other than the oxidation of ^{14}C-succinate derived from 2-keto-^{14}C-glutarate is responsible for the production of ^{14}C-labeled oxaloacetate. The most obvious is evidently a direct carboxylation of pyruvate. The results of McMillan and Mortensen (1963) concerning the metabolism of brain pyruvate imply considerable conversion of pyruvate to oxaloacetate as indicated by the labeling in carbons 2 and 3 of glutamate after pyruvate-2-^{14}C administration. Berl *et al.* (1962) have also shown that the cat brain can fix CO_2 to a significant extent. This is also true in the isolated ventral nerve chain of the

---

[1]E.N. refers to *Enzyme nomenclature* (1964) while E.C. refers to the *Report on the Enzyme Commission* (1961).

lobster where an appreciable fixation of $^{14}CO_2$ occurs in taurine, aspartic, and glutamic acids.

As shown by the results of Table 7-4, the specific activity of aspartate and of taurine are very close to each other but significantly higher than that of glutamate. This finding is consistent with the production of oxaloacetate by carboxylation of pyruvate (Schoffeniels, 1968).

Waelsch and his colleagues have reported analogous results. The fixation of CO_2 in the nerve of bullfrog, rabbit, and lobster has been studied in a variety of experimental conditions. Their results show that besides a fixation at the level of pyruvate, 2-ketoglutarate may also serve as acceptor, thus driving backward the reactions to citrate (Côté et al., 1966; Naruse et al., 1966 a,b). The carboxylation is enhanced during the electrical stimulation or when veratrine is added to the incubation medium. The same effect is observed when the potassium concentration is increased or in a low calcium medium (Cheng and Mela, 1966a,b).

The reality of the carboxylation of pyruvate is unequivocably demonstrated since a pyruvate carboxylase (E.N. 6.4.1.1) has been evidenced in the ox brain (Felicioli et al., 1967).

Table 7-5 shows unpublished results of Gilles and Schoffeniels concerning the fate of ^{14}C-labeled glucose and pyruvate in the ventral nerve chain of the lobster. The specific activity of some amino acids as well as that of taurine are given.

The results obtained with glucose-U-^{14}C show, as expected, that the hydrocarbon skeleton of the labeled amino acids are derived either from pyruvate (Ala, Ser, and Gly) or intermediaries of the Krebs cycle (Glu and Asp). Moreover the values of the specific activities are in the order foreseen by the metabolic sequences involved.

An interesting observation is that the specific activity of taurine is very close to that of alanine. In the lobster nervous system, taurine together with aspartate, glycine, and proline is the most concentrated amino derivative. Its concentration varies between 0.5 to 2 μmoles per 100 mg of fresh weight.

In 1955, Weimberg and Doudoroff have shown that *Pseudomonas saccharophila* may synthesize 2-ketoglutarate from L-arabinose without intervention of the Krebs cycle or the formation

TABLE 7-4

FIXATION OF $^{14}CO_2$ IN THE ISOLATED VENTRAL CHAIN OF THE LOBSTER[a]

Components tagged	Specific activity[b] ($\times 10^{-5}$)
Asp	4.5
Tau	4.1
Glu	3.0

[a] Gilles and Schoffeniels, unpublished data.
[b] Mean of 5 experiments.

of phosphorylated intermediates. These results have been extended by showing that 2-keto-3-deoxy-arabonic acid is an intermediary substrate (Palleroni and Doudoroff, 1956). Datta and Katznelson (1957) have demonstrated that *Acetobacter melagenum* is able to metabolize glucose into 2-ketoglutarate through a pathway involving as intermediaries gluconic, 2-ketogluconic, and 2,5-diketogluconic acids.

That D-arabonate may be a precursor of 2-ketoglutarate in pluricellular organisms is well demonstrated by the results of Gilles and Schoffeniels (Table 7-6).

In the experiments reported, D-arabinose-U-^{14}C was added to the incubation medium of an isolated ventral nerve chain of lob-

TABLE 7-5

SPECIFIC ACTIVITY OF TAURINE AND AMINO ACIDS FOUND IN THE
VENTRAL NERVE CHAIN OF THE LOBSTER AFTER INCUBATION IN THE
PRESENCE OF ^{14}C-LABELED GLUCOSE AND PYRUVATE[a]

Taurine and amino acids	Specific activity ($\times 10^{-3}$) in presence of:					
	Glucose-U-^{14}C			Pyruvate-1-^{14}C		
Ala	7.1	11.0	9.6	32	12.3	15.5
Glu	0.66	2.1	1.7	trace	0.860	1.16
Asp	0.34	0.53	0.60	0.5	0.183	0.5
Tau	11.3	12.0	6.0	5.3	13.4	4.65
Ser	0.55	1.46	3.16	—	1.5	—
Gly	—	0.31	0.09	0.24	0.1	0.245

[a] Gilles and Schoffeniels, unpublished data.

TABLE 7-6
ISOLATED VENTRAL NERVE CHAIN OF THE LOBSTER[a]

| Components tagged | Specific activity ($\times 10^{-3}$) after 1 hour incubation in the presence of arabinose-U-^{14}C | | | | |
	Expt. 1	Expt. 2	Expt. 3	Expt. 4	Expt. 5
Tau	19.0	18.0	36.0	13.0	20.0
Asp	3.0	2.8	2.2	3.6	0.237
Glu	9.7	9.9	2.7	44.0	8.1

[a] After Gilles and Schoffeniels, unpublished results.

ster. The analysis of the amino acids content revealed that the activity was located on taurine, aspartate, and glutamate. This result is interesting since it demonstrates that 2-ketoglutarate may be produced outside the operation of the Krebs cycle (Fig. 7-3); it also indicates that a metabolite of this sequence is common to both taurine and 2-ketoglutarate.

Finally, a decarboxylation of aspartate has been demonstrated in cellular extracts of lobster and crayfish (Gilles and Schoffeniels, 1966) which points to a new metabolic pathway leading to the production of alanine. Together with the demonstration of

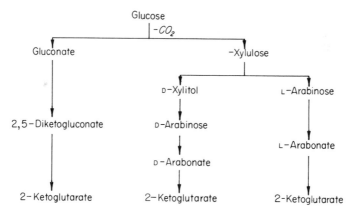

FIG. 7-3. Synthesis of 2-ketoglutarate through pathways outside the Krebs cycle.

the presence of glutamate dehydrogenase and various aminotransferases, the above results suggest that in Crustacea, three of
the most important amino acids in the osmoregulation of the
cells are synthesized along the metabolic route summarized in
Fig. 7-4.

If we compare the effect of veratrine on the metabolism of
amino acids in a stenohaline species, the lobster, and in a euryhaline species, the crayfish, we see that in the former species the
concentrations are not affected, while the turnover rate of aspartate, glutamate, alanine, and serine increases very much. In the
crayfish, on the other hand, the concentration of most of the
amino acids identified increases, except in the case of aspartate
(Table 7-7). This is also true when NH_4Cl is added to saline
bathing the isolated tissue (Table 7-8, 7-9).

An important conclusion to be drawn from these results is that,
in conditions of increased synthesis of amino acids, a stenohaline
species responds by an increase in the catabolism, thus keeping
the intracellular concentration of amino acids nicely balanced.
This phenomenon seems to be an essential feature differentiat-
ing a stenohaline species from a euryhaline one (Gilles and
Schoffeniels, 1964).

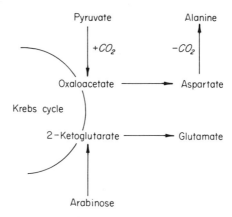

FIG. 7-4. Schematic representation of the metabolic pathway leading to the
synthesis of alanine, glutamate, and aspartate in crustacea. (After Gilles and
Schoffeniels, unpublished data.)

TABLE 7-7

EFFECT OF VERATRINE ($4\mu M$) ON THE CONCENTRATION AND RADIOACTIVITY OF THE AMINO ACIDS FOUND IN THE VENTRAL NERVE CHAIN OF THE LOBSTER AND OF THE CRAYFISH INCUBATED WITH GLUCOSE-U-^{14}C[a]

Amino acids	Lobster				Crayfish	
	Control		Veratrine		Control	Veratrine
	Concentration (μmoles per 100 mg tissue fresh weight)	Radioactivity (cpm per 100 mg tissue fresh weight)	Concentration (μmoles per 100 mg tissue fresh weight)	Radioactivity (cpm per 100 mg tissue fresh weight)	Concentration (μmoles per 100 mg tissue fresh weight)	Concentration (μmoles per 100 mg tissue fresh weight)
Asp	2.205	758.78	2.399	4,017.79	0.510	0.379
Thr	0.069	–	0.067	–	0.045	tr.
Ser	0.572	316.15	0.553	2,705.22	0.241	0.242
Glu	0.534	351.28	0.551	1,668.52	0.209	0.292
Pro	0.833	–	0.848	–	0.064	0.098
Gly	1.884	tr.	1.833	240.26	0.145	0.192
Ala	0.915	6,484.77	0.911	9,704.11	0.239	0.393
Cys	0.083	–	0.087	–	tr.	tr.
Val	–	–	–	–	0.032	0.040
Met	0.006	–	–	–	tr.	tr.
Ileu	0.050	–	0.048	–	0.021	0.027
Leu	0.039	–	0.034	–	0.020	0.034
Tyr	0.035	–	0.037	–	0.026	0.035
Phe	0.027	–	–	–	0.023	0.027
Ammonia	0.493	–	0.455	–	–	–
Lys	0.031	–	0.028	–	–	–
His	0.049	–	0.044	–	–	–
Arg	0.320	–	0.317	–	–	–

[a]After Gilles and Schoffeniels, unpublished data.

129

TABLE 7-8

EFFECT OF AMMONIUM CHLORIDE ON THE CONCENTRATION AND
RADIOACTIVITY OF THE AMINO ACIDS FOUND IN THE VENTRAL NERVE CHAIN
OF THE LOBSTER INCUBATED WITH PYRUVATE-1-^{14}C[a]

Amino acids	Control		Ammonium chloride	
	Concentration (μmoles per 100 mg tissue fresh weight)	Radioactivity (cpm/100mg tissue fresh weight)	Concentration (μmoles per 100 mg tissue fresh weight)	Radioactivity (cpm/100 mg tissue fresh weight)
Asp	3.231	591	2.969	2273
Thr	0.172	—	0.196	—
Ser	0.657	986	0.663	1017
Glu	0.976	845	0.992	1249
Pro	1.389	—	1.351	—
Gly	2.055	218	1.992	189
Ala	1.133	13,885	1.143	13,916
Cys	0.084	—	0.087	—
Val	—	—	—	—
Met	—	—	—	—
Ileu	0.053	—	0.059	—
Leu	0.033	—	0.035	—
Tyr	—	—	—	—
Phe	—	—	—	—
Ammonia	0.204	—	0.210	—
Lys	0.036	—	0.036	—
His	0.023	—	0.024	—
Arg	0.488	—	0.499	—

[a] Gilles and Schoffeniels, unpublished data.

Returning to the ecological aspects which are at the origin of our discussion, we may propose that the cationic composition of the cell controls the composition of the intracellular amino acid pool by influencing the production of free amino acids as well as their catabolism.

When a euryhaline invertebrate is transferred from fresh water to seawater, there is, as we have seen, an increase in the concentration of intracellular inorganic constituents, and it may be suggested that, in this case, there is an influence of such a change on the steady state of the amino acid pool. According to this hypothesis, the activity of the enzyme system directly involved in the metabolism of amino acids of the intracellular fluid should be differentially affected by the ionic composition (Schoffeniels, 1960).

To test this hypothesis, a study was made of the effect of ions on the rate of activity of various enzymes. In the following sections, the essential results, some still unpublished, will be summarized (see also Schoffeniels, 1968b).

L-*Glutamate:NAD Oxidoreductase (E.N. 1.4.1.2)*

The enzyme found in animal tissues requires NAD as coenzyme (E.N. 1.4.1.2). However in the liver NADP may also be used (E.N. 1.4.1.4). Enzymatically active glutamate dehydrogenase consists of an association of monomers. When dissociated from each other, a situation brought about by dilution, steroid hormones, etc., the loss of activity toward glutamate is paralleled

TABLE 7-9

EFFECT OF AMMONIUM CHLORIDE ON THE AMINO ACIDS CONCENTRATION IN
THE VENTRAL NERVE CHAIN OF THE LOBSTER AND CRAYFISH[a]
(μMOLES/100 MG TISSUE FRESH WEIGHT)

Amino acids	Crayfish		Lobster	
	Control	NH$_4$Cl $10^{-2} M$	Control	NH$_4$Cl $10^{-2} M$
Asp	0.410	0.279	4.503	5.306
Thr	0.035	tr.	0.185	0.170
Ser	0.141	0.142	1.020	0.885
Glu	0.109	0.192	0.880	0.780
Pro	0.054	0.048	1.033	0.938
Gly	0.105	0.112	2.832	2.700
Ala	0.139	0.393	1.109	1.115
Val	0.022	0.039	0.121	0.112
Met	tr.	tr.	0.022	0.020
Ileu	0.011	0.016	0.063	0.063
Leu	0.025	0.039	0.081	0.070
Tyr	0.016	0.024	tr.	tr.
Phe	0.015	0.019	tr.	tr.

[a] Gilles and Schoffeniels, unpublished data.

by the apparition of catalytic properties for monocarboxylic amino acids. The intact molecule is formed by identical subunits of molecular weight around 53,000. It is worthwhile to recall that in *Neurospora*, glutamate dehydrogenase is also composed of identical subunits. In the case of beef liver cells, the enzyme has been subjected to thorough analysis with regard not only to kinetic properties but also as to the amino acid composition. As a result it has been found that the subunit is a peptide chain, the N-terminal of which is an alanine residue, while the C-terminal belongs to threonine (Appella and Tomkins, 1966).

Kinetic studies indicate that the catalytic activity is enhanced by ADP, some amino acids, and some inorganic cations. On the contrary, the enzyme activity is inhibited by GTP, estrogenic hormones, inorganic anions, etc. A careful analysis of these results suggests that for a molecular weight of 1.1×10^6 the enzyme possesses 20 sites of fixation for the coenzymes, the substrate, ADP, and GTP. Thus one site for each compound is available per subunit.

The enzyme accepts on the same site either NAD^+-NADH or $NADP^+$-NADPH. However at high concentrations NAD is activator of the oxidative deamination of glutamate while the reduced coenzyme inhibits the reverse reaction. This phenomenon is not observed with NADPH. On the basis of these results, Frieden (1959) has suggested the existence of two categories of sites for the fixation of NAD, one directly involved in the catalytic process, the other playing an indirect role by controlling the enzyme activity through the state of aggregation of the subunits.

Recent experiments indicate that the glutamate dehydrogenase has, in the absence of other substrates, two different types of sites fixing NADH. One site is specific to adenosine 5'-diphosphate, ADP or NADH. The other site is specific for the nicotinamide moiety of the coenzyme and thus indifferently accepts NAD or NADP (Iwatsubo *et al.*, 1966).

In the presence of substrate the affinity of the two categories of site changes and kinetic studies of the formation of the various complexes involved in the catalytic steps lead to the conclusion that the interactions with allosteric effectors may variously affect the enzyme activity depending on the conditions of the experimentation.

This makes extremely difficult an extrapolation of the results obtained *in vitro* to the situation found *in vivo*. Moreover most of the kinetic data are obtained with concentrations in either coenzyme or enzyme known to be unrealistic. They are however of great value in showing how intracellular regulation of an important catalytic step could be effected. They also point to the fact that the enzyme activity may be directly related to the state of aggregation of the subunits according to the following scheme:

$$\text{Polymer} \quad \overset{\text{I}}{\rightleftharpoons} \quad \text{monomer x} \quad \overset{\text{II}}{\rightleftharpoons} \quad \text{monomer y}$$

The three forms are catalytically active but the substrate specificity varies. While monomer x is primarily specific to glutamate dehydrogenase reaction, monomer y has enhanced activity toward certain monocarboxylic amino acids (alanine for instance). The three forms are in equilibrium and the various agents known to affect the enzyme activity are thought to displace the equilibrium in one or the other direction. For instance, ADP would favor the formation of monomer x while GTP and estrogenic hormones would shift the equilibrium toward the formation of monomer y. The form x of the monomer has a great tendency to aggregate and is antigenically different from the form y. These findings have not yet received an explanation. They however point to the many possibilities of the cell to control the activity of this enzyme since most of the experimental conditions used could be of biological significance. Of special interest is the fact that the activity of glutamate dehydrogenase (GDH) extracted from the tissues of spiny lobster (*Palinurus vulgaris* Latr.), crayfish (*Astacus fluviatilis* L.), and lobster (*Homarus vulgaris* L.) is dependent on the ionic composition of the incubating medium (Schoffeniels and Gilles, 1963; Schoffeniels, 1964a,b, 1965a, 1966a). In the experiments, the enzyme activity is estimated by measuring at 340 mμ the rate of disappearance of reduced diphosphopyridine nucleotide (NADH) in the presence of 2-ketoglutarate and NH$_4$ ions. In order to evaluate the exact role of various ionic species, the same salt of different cations has been used (Table 7-10). As far as the GDH extracted from lobster muscle is concerned, it can be seen that both the anion and the

TABLE 7-10

GDH EXTRACTED FROM *Homarus vulgaris* L. EFFECT OF VARIOUS ANIONS
AND CATIONS ON THE INITIAL VELOCITY (v_i).

	Enzyme activity	
Salt used	Gills (v_i)	Muscle (v_i)
200 μmoles/ml		
KNO_3	9	–
$NaNO_3$	9	–
NH_4NO_3	16	–
270 μmoles/ml		
KNO_3	–	10 – 10.5
$NaNO_3$	–	11.9 – 13.9
NH_4NO_3	–	18.5
NaAc	–	15.3
KAc	–	18.2
NH_4Ac	–	18.4
NaCl	–	35.7
KCl	–	33.3
NH_4Cl	–	33.3

[a]The enzyme activity is measured in the presence of 50 μmoles/ml of NH_4NO_3 (gills) or 20 μmoles/ml of NH_4 acetate (muscle). After Schoffeniels 1964a.

cation are important in determining the enzyme activity. In the case of chloride ions the type of cation used does not seem to matter much. On the contrary, in the presence of other anions, one observes marked differences between Na, K, and NH_4 cations. This is further established by comparing the oxalate of Li, K, and NH_4 (Table 7-11).

It is evident from the results obtained that there is a rather interesting specificity of the enzyme toward the type of cation used. The results of Table 7-11 acquire even more significance if one considers the enzymatic activity of the two extracts assayed in standard conditions. As a matter of routine, the activity of various extracts is always measured in the same conditions to verify the reproducibility of our extraction procedure. We measure the rate of oxidation of NAD using 100μmoles/ml of NH_4Cl. The standard activity of the extracts used to obtain the results re-

TABLE 7-11

GDH Extracted from *Homarus vulgaris* L. Effect of Various Cations on the Initial Velocity (v_i) Measured in the Presence of 50 μmoles/ml of NH_4NO_3[a]

| | Enzyme activity | |
Salt used	Gills (v_i)	Muscle (v_i)
A. 150 μmoles/ml		
Li oxalate	5	12
K oxalate	24	18
NH_4 oxalate	21	60
B. 210 μmoles/ml		
$LiNO_3$	19	2
$NaNO_3$	32	5
NH_4NO_3	35	5

[a]After Schoffeniels, 1964a.

ported in Table 7-11, A, is given in Table 7-12. In the same conditions of assay, the gills extract is about twice as active as the extract from muscles.

It is clear from the above results that, in the lobster, GDH extracted from muscle and gills is sensitive to the ionic composition of the incubating medium. Both the anion and the cation are

TABLE 7-12

GDH Extracted from *Homarus vulgaris* L. Standard Conditions of Assay (100 μmoles/ml NH_4Cl)[a]

| | Enzyme activity | | | |
| | Gills NaCl μmoles/ml | | Muscle NaCl μmoles/ml | |
	Expt. 1	Expt. 2	Expt. 1	Expt. 2
v_i Initial velocity	0	400	0	400
	1	76	5	44

[a]After Schoffeniels, 1964a.

important since it is possible to demonstrate a variable efficiency according to the ionic species used. It is interesting to note that the specificity of the cation can only be demonstrated with certain anionic species. This would suggest the existence of two categories of specific sites at the surface of the enzyme. These sites may be saturated by anions and cations and, according to the species used, a variable activation of the enzyme is obtained. Thus both categories of sites would contribute to the overall enzyme activity. This could explain why the cation specificity shows up only in the presence of certain anionic species.

It is also worth commenting on the fact that GDH extracted from muscle does not behave exactly like the enzyme extracted from the gills. The cationic specificity, as well as the anionic efficiency, are different according to the origin of the enzyme. This most likely corresponds to differences in the structure of the enzyme proteins and offers a new example of isoenzyme.

If we now turn to the results obtained with the crayfish tissues it has been shown that the activity of GDH is also ion sensitive. But contrary to what is observed with *Palinurus* or *Homarus*, even in the absence of 2-ketoglutarate and NH_4 ions, NAD is oxidized by the tissue extract (gills or muscle). This process is also sensitive to the ionic composition of the incubating medium (Schoffeniels, 1965a), but the initial velocity decreases rather markedly with time. We may wonder if the oxidation of NAD does not require molecular oxygen as electron acceptor. This is not the case since in the presence of nitrogen or sodium cyanide (1 mM) the initial velocity remains equal to that of the control.

However, after dialysis, the extract loses its capability of oxidizing NAD, while GDH activity is unchanged. If the dialyzate is added to the extract, the activity is restored. The necessary factor is thermostable and when added at increasing concentration coenzyme oxidation follows the Michaelis-Menten kinetics. Thus it seems likely that a substrate requiring NAD is present in the extract. Pyruvate and oxaloacetate can be excluded on the basis of experimental results showing that the ionic requirements of lactic and malic dehydrogenases are different from those of the system oxidizing NAD.

An effect of the ionic composition of the incubating medium may also be demonstrated when using an enzyme preparation of

TABLE 7-13

RELATIVE ACTIVITY OF GDH (CRYSTALLIZED FROM
BEEF LIVER) AS A FUNCTION OF pH AND TYPE OF BUFFER USED[a]

Buffer	pH	C	NaCl	NaNO$_3$	KCl	KNO$_3$	NH$_4$Cl	LiCl
Histidine sucrose	7	80	105	76	80	80	164	38
(15 μmoles/ml)	7.3	100	190	115	115	85	145	58
	7.8	120	200	115	140	95	225	67
Tris acetate	7	60	100	120	70	110	170	30
(20 μmoles/ml)	7.3	100	140	140	100	143	250	67
	7.8	60	180	180	90	100	180	60
Tris Cl	7	85	145	110	70	95	190	30
(20 μmoles/ml)	7.3	100	167	133	117	133	200	33
	7.8	85	450	294	190	170	260	55
PO$_4$	7	100	90	100	55	120	135	20
(11 μmoles/ml)	7.3	100	110	85	55	120	120	20
	7.8	80	105	50	50	80	135	20

[a] The activity of the control at pH 7.3 is considered to be equal to 100. The enzyme concentration is 1 μg/3 ml reaction mixture, except in the case of phosphate buffer where it is 8 μg. After Schoffeniels, 1966b.

commercial origin (Table 7-13). It can be seen that the anion as well as the cation play an important role. The kinetic studies of the reaction have shown that the anions are inhibitors of the enzyme activity by competing with 2-ketoglutarate for receptor sites on the enzyme (Fig. 7-5).

In order of decreasing effectiveness, nitrate, acetate, and chloride are found. Among the cations tested, lithium is the only one to be inhibitory over the entire range of concentrations assayed. The other cations are activators possibly by favoring the equilibrium toward the formation of monomer x and its state of aggregation.

There is an optimal concentration in cations above which they are inhibitory as indicated by the results of Fig. 7-6. This is also observed in the crude cellular extracts obtained from various sources (Schoffeniels, 1965a, 1966a). The optimal concentration is however different according to the origin of the material. In mammals and freshwater species it is generally around 50 μ

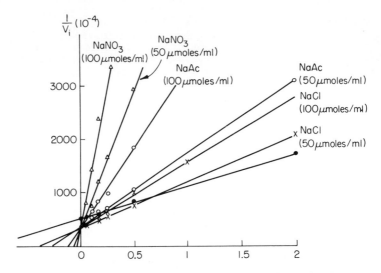

FIG. 7-5. Glutamic dehydrogenase. Plot of reciprocal initial velocity versus reciprocal 2-ketoglutarate concentration (μmoles/ml). Effect of three sodium salts. (Schoffeniels, 1966b.)

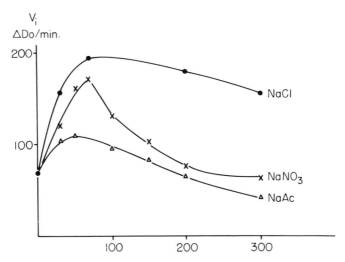

FIG. 7-6. Glutamic dehydrogenase. Initial velocity as a function of salt concentration (μmoles/ml) in the case of three different anions. (Schoffeniels, 1966b.)

moles NaCl/ml while in the case of marine species it is well above that value and situated in the neighborhood of 500 μmoles NaCl/ml. Since the fixation of the reduced dinucleotide on the enzyme is accompanied by an increase in the fluorescence of the coenzyme, a way of studying quantitatively the factors influencing the formation of the complex is obtained.

It has been shown that experimental conditions known to decrease the catalytic properties of the enzyme increase the number of sites fixing the reduced coenzyme. The reverse is true when using ADP, an agent known to increase the enzyme activity.

The inorganic ions have the same effect in that they decrease the number of binding sites for NADH (Duyckaerts and Schofeniels, 1966). Since in the same conditions the catalytic activity of the enzyme increase, it can be concluded that the sites involved are not catalytically important. They could also be related to the sites involved in the association of the monomers or be allosteric sites controlling the transconformation of the polypeptide chain.

By considering the standard oxidoreduction potentials (E'_o) of the systems involved, it is apparent that the enzyme favors the reductive amination rather than the reverse reaction. The E'_o values for the systems NADH/NAD and glutamate/2-ketoglutarate are respectively −0.32 V and −0.108 V indicating that the presence of any concentration of ammonia or 2-ketoglutarate above the infinitesimal would strongly oppose the oxidative deamination of glutamate. This conclusion is of great practical consequence, since it indicates that the enzyme must play an important role in the nitrogen metabolism by controlling the entrance of ammonia, the inorganic form of nitrogen commonly used in animal cells, in organic compounds.

This key function, already noticed by Dewan (1938) and von Euler et al. (1938) many years ago, has been overlooked by many authors, especially renal physiologists who insisted on the importance of the enzyme in the production of ammonia (see also Hird and Marginson, 1966). The importance of the glutamate dehydrogenase in controlling the synthesis of glutamate has now been recognized by many authors (Olson and Anfinsen, 1953; Strecker, 1953; Weil-Malherbe, 1957; Klingenberg and Pette,

1962; Tager and Slater, 1963a; Schoffeniels, 1964a, 1965a, 1967a,b).

The results presented above point to a possible role of the inorganic composition of the cell in controlling the activity of this enzyme, a suggestion that may be of special significance when considering the adaptation to media of various salinities. Thus a euryhaline crustacea, when introduced into brackish water or seawater, undergoes an increase of the inorganic effectors of the cell. This change exerts an action on the activity of glutamate dehydrogenase. Consequently the intracellular concentration of glutamate increases, thus contributing via the aminotransferases to the isosmotic intracellular regulation.

This concept is in accordance with the observation mentioned earlier in the text showing that when *Carcinus maenas* or *Eriocheir sinensis* are transferred into diluted medium, the nitrogen excretion is increased for a certain period of time, while the opposite effect is observed when the crab is transferred into seawater (Needham, 1957; Jeuniaux and Florkin, 1961; Florkin *et al.*, 1964) (Fig.7-1 and 7-2).

Lactic Dehydrogenase (E.N. 1.1.1.27)

The lactic dehydrogenase (LDH) catalyzes the reduction of pyruvate into lactate

$$\text{Pyruvate} + \text{NADH} + \text{H}^+ \to \text{Lactate} + \text{NAD}$$

The oxidoreduction potentials of the couples lactate/pyruvate and NADH/NAD are such that the formation of lactate is favored. Various forms of catalytically active proteins have been identified and it is well demonstrated that the enzyme exists in five main forms resulting from the association of two different monomers H and M (Kaplan, 1965). The monomers have a molecular weight around 35,000 and the enzyme results from a tetrameric association giving rise to five possible forms (H_4, H_3M_1, H_2M_2, H_1M_3, M_4) that can be separated by electrophoresis. They have basis of genetic studies as well as amino acid analysis, arguments have been presented favoring the idea that the synthesis of the polypeptides H and M is controlled by two different genes. H_4-LDH is maximally active at low pyruvate concentration and is inhibited by a substrate concentration of 10^{-2} M while the M_4-LDH is active at high pyruvate concentration. The activity of the M_4-LDH is also higher than that of the H_4-LDH when the 3-acetylpyridine analog of NAD is used. Accordingly a comparison of relative rates of reaction with two concentrations of pyruvate or with several coenzyme analogs may allow the H- and M-LDH's of a species to be easily distinguished.

The physiological significance of the various forms of LDH is not fully understood. A positive correlation may be found between the presence of the H-type in a given cell and the availability of oxygen (see for instance Salthe, 1965) which points to the fact that the H form of the enzyme is associated with a metabolism predominantly aerobic. This obviously makes sense since the H_4-LDH is inhibited at high pyruvate concentration. It is also worth noting that the LDH found in the mitochondria is of the H-type (Agostoni *et al.* 1966). The proportion of the hybrids of LDH may however vary according to the conditions of the surroundings. Thus a change in oxygen tension, the denervation, and some pathological states or various hormones may affect profoundly the structure of the LDH found intracellularly. It is generally assumed that this is the result of a differential effect on the synthesis of the monomers M and H. Estradiol favors the synthesis of the M polypeptide while a high oxygen pressure inhibits its synthesis. Oxygen also enhances the synthesis of the H polypeptide (Dawson *et al.*, 1964). There is however little information available as to the factors that may influence the tetrameric association and the formation of hybrids.

In crustacea there appear to be two types of LDH, and the use of coenzyme analogs lead to the conclusion that in the muscles of the species studied the form M of the enzyme is dominant (Kaplan *et al.*, 1960). An interesting observation related to our problem of the control of the intracellular hydrogen transport is that the activity of the enzyme, partially purified by ammonium sulfate fractionation, is inhibited by sodium chloride. This is true for all the species we have studied so far, including mammals (Schoffeniels, 1968a,b and Table 7-14).

However if the same determination is performed at a concentration of pyruvate of $10^{-2}M$, some interesting differences are observed (Table 7-15). In the case of the two euryhaline species

TABLE 7-14

Muscle Lactic Dehydrogenase of Various Crustacea as a Function
of the NaCl Concentration in the Incubating Medium[a]

Crustacean	Enzyme at salt concentration (mM)									
	0	50	100	150	200	250	300	400	500	600
Maia squinado	100	100	70	60	36	28	8	8	—	—
Carcinus moenas	100	105	75	60	35	35	25	10	—	—
Cancer pagurus	100	75	50	23	30	12	9	5	—	—
(Big specimen)	100	—	38.5	—	12	—	6.9	5.1	3.3	2.3
(Small specimen)	100	—	45	—	12	—	5.5	5.1	2.0	2.8
Portunus puber	100	125	120	80	62	50	35	10	—	—
Astacus fluviatilis	100	94	66	56	47	34	34	28	—	—
Homarus vulgaris	100	—	58	—	12	—	5.1	4.1	4.1	5.1
Eriocheir sinensis	100	—	67	—	20	—	10.5	6.7	5.5	4.5

[a]The enzyme is partially purified by ammonium sulfate precipitation between
30% and 70% of saturation and the pyruvate concentration in the assay mixture
is 10^{-4} M. Relative activity. Schoffeniels, 1968b.

TABLE 7-15

Muscle Lactic Dehydrogenase of Three Crustacea as a Function
of Pyruvate and NaCl Concentration[a]

Crustacean	Enzyme activity measured in various NaCl concentrations (mM)						
	0	50	100	200	300	400	500
Pyruvate : 5×10^{-4} M							
Astacus fluviatilis	100	93	71	57	36	29	14
Eriocheir sinensis	100	121	121	89	44	33	22
Homarus vulgaris	100	105	100	90	80	75	65
Pyruvate : 10^{-2} M							
Astacus fluviatilis	100	100	100	78	78	68	71
Eriocheir sinensis	100	103	100	93	77	70	57
Homarus vulgaris	100	125	125	125	125	125	125

[a]Schoffeniels, 1968b.

[b]The enzyme of *Astacus* is partially purified (see legend Table 7-14) while
those of *Eriocheir* and *Homarus* are crude cellular extracts in distilled water.

(*Astacus* and *Eriocheir*), high concentrations of NaCl are inhibitory whatever the concentration in pyruvate, while the enzyme of *Homarus*, a stenohaline species, is not inhibited by NaCl if the pyruvate concentration is high. It is thus clear that in the case of the two euryhaline species at concentrations in NaCl rather close to what is found in the cell of a marine species, there is less synthesis of lactate, thus making more reducing equivalents available for other purposes, e.g., glutamic acid synthesis.

3-Glycerophosphate Dehydrogenase (E.N.1.1.1.8)

The NAD-dependent glycerol-3-phosphate dehydrogenase is located in the cytoplasm whereas the cytochrome-linked enzyme is located in the mitochondria. As mentioned earlier in the text (see Chapter IV, p. 57), this enzyme catalyzes the reduction of phosphodihydroxyacetone into 3-phosphoglycerol which permits the transfer of reducing equivalents originating in the cytoplasm to the respiratory chain in the mitochondria. Its activity is related to the ionic composition of the incubation medium and some differences exist when one considers the enzyme extracted from euryhaline or stenohaline species (Table 7-16). At high concentration in NaCl the enzyme activity is inhibited. However for values of NaCl between 100 and 200 mM, the activity is very much enhanced when considering the euryhaline species.

Glyoxylate Reductase (E.N.1.1.1.26)

This enzyme catalyzes the reduction of glyoxylate to glycollate. It also reduces hydroxypyruvate to D-glycerate. Its metabolic importance is far from being understood but it certainly plays a role in the regulation of glycine concentration inside the cell. The enzyme extracted from muscle and nerve of crustacea is inhibited by inorganic ions (Table 7-17).

Isocitrate Dehydrogenase

Two enzymes are of metabolic importance, one is located in the mitochondrion (E.N. 1.1.1.41) where it plays an important

TABLE 7-16

MUSCLE 3-GLYCEROPHOSPHATE DEHYDROGENASE OF SOME CRUSTACEA[a]

Crustacean	Enzyme activity[b] with various NaCl concentrations (mM)						
	0	100	200	300	400	500	600
Euryhaline							
Eriocheir sinensis	100	400	360	280	200	–	–
	100	300	280	240	160	120	100
Carcinus moenas	100	305	283	250	200	133	100
Portunus puber	100	212	225	206	169	131	94
Astacus fluviatilis	100	366	315	217	195	183	116
Stenohaline							
Homarus vulgaris	100	145	117	93	50	42	33
	100	155	135	105	80	–	–
Maia squinado	100	180	133	126	93	–	–
	100	158	141	125	91.5	58.5	45
Cancer pagurus							
big specimen	100	180	120	120	120	–	–
small specimen	100	126	175	126	75	–	–

[a]Schoffeniels, unpublished data.
[b]The enzyme is partially purified by ammonium sulfate precipitation between 30 and 70% saturation and the concentration in dihydroxyacetone phosphate is $5 \times 10^{-4} M$. Relative activity.

TABLE 7-17

MUSCLE GLYOXYLATE REDUCTASE OF SOME CRUSTACEA[a]

Crustacean	Enzyme activity[b] with different NaCl concentrations (mM)						
	0	100	200	300	400	500	600
Homarus vulgaris	390	165	95	60	40	20	25
Cancer pagurus (big specimen)	1000	160	70	40	30	25	25
Eriocheir sinensis	680	620	95	60	40	25	20
Carcinus moenas	600	185	95	45	35	25	20

[a]Schoffeniels, 1968b.
[b]The enzyme is partially purified by ammonium sulfate precipitation between 30 and 70% saturation. Initial velocity expressed in OD units per minute.

role in the operation of the tricarboxylic acid cycle. It involves NAD as hydrogen acceptor (Ernster and Navazio, 1957; Kaplan *et al.*, 1956). The other (E.N. 1.1.1.42), extremely concentrated in many cells, is found in the cytoplasm; it is NADP-linked. An interesting observation is that the enzyme activity is related to the ionic composition of the medium. This is true with a crude cellular extract or after partial purification. The soluble enzyme is precipitated by ammonium sulfate between 30 and 70 % saturation.

This enzyme is found in the muscle and nervous system of marine invertebrates. It is however more concentrated (or more soluble) in the euryhaline species than in the stenohaline species. In both types of organism the activity decreases as the salt concentration increases. This has been observed by Gilles and Schoffeniels (unpublished) not only with crustacea but also with mollusks (Table 7-18).

TABLE 7-18

ISOCITRATE DEHYDROGENASE EXTRACTED FROM LOBSTER AND
CRAYFISH VENTRAL NERVE CHAIN[a]

Crustacean and salt	Enzyme activity[b] with various NaCl concentrations (mM)			
	0	100	200	400
Homarus vulgaris				
NaCl	1	0.75	0.61	0.35
	1	0.81	0.68	0.40
KCl	1	0.74	0.59	0.35
	1	0.80	0.66	0.39
Astacus fluviatilis				
NaCl	1	0.74	0.50	0.23
	1	0.69	0.45	0.18
KCl	1	0.72	0.51	0.25
	1	0.67	0.44	0.19

[a]After Gilles and Schoffeniels, unpublished results.
[b]The results are expressed in relation to the control value taken as 1.

L-*Malate:NADP Oxidoreductase (E.N. 1.1.1.40)*

This enzyme, catalyzing reversibly the carboxylation of pyruvate, was formerly called malic enzyme. Its activity is estimated by measuring the reduction of NADP at 340 mμ. In the presence of added NaCl or KCl the activity of the enzyme decreases (Table 7-19).

L-*Malate:NAD Oxidoreductase (E.N. 1.1.1.37)*

Malate dehydrogenase catalyzes the reduction of oxaloacetate into malate. The activity of this enzyme is thus estimated by measuring at 340 mμ the oxidation rate of reduced NAD. As the standard oxidoreduction potentials of the system NADH/NAD is higher (-0.32 V) than that of the system malate/oxaloacetate (-0.17 V), it is obvious that the reduction of oxaloacetate is very much favored. In the presence of increasing amounts of NaCl or KCl the enzyme activity is slightly stimulated rather than being inhibited (Table 7-20). It is interesting to note that the optimum concentration of salt is much lower in the case of the crayfish, a freshwater species, than in the case of lobster.

TABLE 7-19

MALIC ENZYME (E.N. 1.1.1.40) EXTRACTED FROM LOBSTER
AND CRAYFISH NERVE CHAIN[a]

	Enzyme activity at salt concentration (mM):				
Crustacean and salt	0	100	200	400	600
Homarus vulgaris					
NaCl	100	72	52	38	34
KCl	100	69	53	39	35
Astacus fluviatilis					
NaCl	100	56	37.5	24	17
KCl	100	63	38	26	15

[a]After Gilles and Schoffeniels. Crude cellular extract in distilled water. Relative values.

TABLE 7-20
MALATE DEHYDROGENASE (E. N. 1.1.1.37) EXTRACTED
FROM LOBSTER AND CRAYFISH MUSCLES[a]

Crustacean and salt	Enzyme activity at salt concentration (mM)									
	0	25	50	75	100	150	200	400	600	800
Homarus vulgaris										
NaCl	100	105	110	—	130	—	155	161	119	97
KCl	100	105	117	—	133	—	152	161	117	92
Astacus fluviatilis										
NaCl	100	—	106	119	106	94	87	—	69	—
KCl	100	—	112	125	109	91	81	—	62	—

[a] After Gilles and Schoffeniels, unpublished results. Crude cellular extract in distilled water. Relative values.

L-*Malate Hydro-lyase (E.N. 4.2.1.2)*

Fumarate hydratase catalyzes the formation of malate from fumarate and H_2O. Its activity is estimated by measuring at 240 mμ the disappearance of fumarate. As shown in Table 7-21, NaCl has a different effect whether one considers the enzyme extracted from the cells of a euryhaline or a stenohaline species. It is indeed apparent that in the case of the lobster, as the salt concentration increases, the enzyme activity goes through a maximum and then decreases, while in the case of the crayfish the results show a progressive inhibition. Their behavior seems to be related to the adaptive abilities of the species to media of different salinities and thus provide a partial explanation of the euryhalinity at the molecular scale of dimensions.

To conclude, it may be proposed that the inorganic composition of the intracellular fluid has a marked influence on the activity of enzymes controlling key reactions in the cell metabolism. Our results suggest that in the euryhaline species the glutamate synthesis may be considered as an important factor controlling the activity of the respiratory chain. This control does not seem to exist in the stenohaline species. The activity of key enzymes such as the lactic or the 3-phosphoglycerol dehydrogenases is

TABLE 7-21

FUMARATE HYDRATASE FROM THE MUSCLES OF LOBSTER AND CRAYFISH[a]

Crustacean activity	Enzyme activity at NaCl concentration (mM)							
	0	10	25	50	100	150	200	400
Homarus vulgaris	100	–	–	200	–	180	160	80
	100	–	107	115	92	–	45	–
Astacus fluviatilis	100	97	91	87	61	–	37	–
	100	91	82	74	58	–	40	–

[a]Gilles and Schoffeniels, unpublished results. Crude cellular extract in distilled water. Relative values.

differently affected by the inorganic ions according to their zoological origin.

The interpretation of the role of glutamic dehydrogenase in controlling the pool of ammonia within the cell has been presented a few years ago (Schoffeniels, 1964b). It finds an experimental support in the results obtained by Péquin and Serfaty (1966). These authors have shown that an effective decrease in ammonia excretion by gills results from the infusion of glutamate in the carp, *Cyprinus carpio*.

It is obvious from the above results that the activity of many enzymes is influenced by the ionic composition of the incubating medium. That we are dealing with a specific effect is demonstrated by the kinetic analysis of the results and by the fact that some enzymes such as L-aspartate: 2-ketoglutarate aminotransferase or aspartate decarboxylase are unaffected by the ions. It is also interesting to notice that in the conditions of high concentration of salts, the activity of some of the key enzymes involved in the transfer of hydrogen is modified in such a way that the cell metabolism is geared toward anaerobic conditions, which prevents some of the reducing equivalents from entering the respiratory chain and making them available for the synthesis of glutamate for instance. Moreover the kinetic analysis of the effects of inorganic ions in the activity of these enzymes render it possible to detect subtle evolutionary changes in a given enzyme.

As is well known, the synthesis of glutamate from its corresponding keto acid is dependent on the availability of reducing equivalent. In that respect, one may thus consider that a competition must exist between the various synthetic sequences needing NAD or NADP. As a matter of consequence one should find some kind of interaction between the respiratory chain and the synthesis of glutamate. This is indeed the case as shown by the results of oxygen consumption in species known to synthesize amino acids at various rates according to the nature of their environment. The oxygen consumption in most of the euryhaline species so far studied is increased on transfer from seawater to diluted seawater (see Gilles and Schoffeniels, 1965; Kii_g, 1965, and Table 7-22).

In the spider crabs, *Maia* and *Libinia,* which do not normally invade brackish water, respiration is significantly depressed when the animal is placed in dilute seawater (see also Schwabe, 1933). In contrast, for *Carcinus* and *Callinectes* which normally invade brackish water, a considerable increase in oxygen con-

TABLE 7-22
OXYGEN CONSUMPTION OF WHOLE ANIMALS
IN RESPONSE TO DILUTED SEAWATER[a]

Animal[b]	μl O_2/hour/gm in seawater	μl O_2/hour/gm in 50% seawater	Ratio $\dfrac{\mu\text{l } O_2 \text{ in 50\% seawater}}{\mu\text{l } O_2 \text{ in 100\% seawater}}$
Maia (5)	54.2±4.4	28.2±4.6	0.52±0.04 ($P<0.005$)
Libinia (6)	37.1±3.8	26.0±2.8	0.70±0.03 ($P<0.001$)
Carcinus (14)	93.6±6.2	127.5±9.5	1.33±0.04 ($P<0.001$)
Callinectes (6)[c]	97.3±7.5	150.4±13.8	1.53±0.05 ($P<0.001$)

[a]After King, 1965.

[b]The number of animals is in parentheses. Temperature is 25°C.

[c]For *Callinectes,* the diluted seawater was 20% and the ratio indicates the response at this dilution to that in full-strength water during "control" period.

sumption is observed. The results for *Carcinus*, showing a 33 % increase in 50 % seawater as compared to that in normal seawater are in agreement with those obtained by Schlieper (1929) and Schwabe (1933). These observations are best explained if one postulates that the reduced dinucleotide coenzyme diverted from the glutamate synthesis is oxidized via the respiratory chain in the case of a euryhaline species [4]. The decrease in oxygen consumption observed when dealing with a stenohaline species may result from agonic reactions since the animals did not survive more than a few hours in the experimental conditions. Other explanations could be offered and the decrease in the oxygen consumption could have more significance since a euryhaline freshwater crab such as *Paratelphusa hydrodromous* shows minimal oxygen consumption in 50 % seawater and maximal values in tap water and in 100 % seawater (Ramamurthi, 1967).

In our opinion glutamate dehydrogenase plays a role as important as that generally devoted to lactic dehydrogenase of the respiratory chain in controlling the oxidoreduction state of the cell dinucleotide coenzymes (Schoffeniels, 1968a, b). The two major sources of reducing power are derived from the activity of the glycolysis together with that of the tricarboxylic acid cycle. As the mitochondrial membrane is impervious to the reduced dinucleotides, the cell must have at its disposal a means of passing the electrons produced in the cytoplasm into the mitochondrion.

As a result of the activity of the oxidation step leading to 1,3-diphosphoglycerate, NAD^+ in the extramitochondrial cytoplasm is continuously reduced. Other extramitochondrial dehydrogenases may also contribute to this phenomenon. Despite the inability of NADH to cross the mitochondrial membrane, electrons of cytoplasmic NADH are able to reach the respiratory chain through the so-called shuttle system. Two mechanisms are generally proposed: the acetoacetate and the dihydroxyacetone phosphate shuttles (Devlin and Bedell, 1960; Boxer and Devlin, 1961).

[4]In *Eriocheir* however the respiratory rate remains constant (Schwabe, 1933). As discussed below this can be explained if one postulates that glutamate synthesis acts as an uncoupling agent on the oxidative phosphorylation.

As a result of the functioning of the shuttles the reducing equivalents of the extramitochondrial NADH are passed into the mitochondrion while the molecules of NAD are resynthesized, thus permitting the glycolysis to proceed.

The extramitochondrial ratio NADH/NAD is also controlled by the activity of various dehydrogenases. Generally the lactate and alcoholic dehydrogenases are considered to be the most important. The activity of these two enzymes explains indeed why, in anaerobic conditions, glycolysis still proceeds with the unescapable consequence of lactate and/or ethanol production. The ratio of lactate (or ethanol)/pyruvate is thus a measure of the oxidoreduction state of the glycolytic NAD.

Other synthetic systems involving the reduced dinucleotide coenzymes are undoubtedly concerned in the regulation of the oxidoreduction state of glycolytic and respiratory coenzymes. When discussing the integration of the enzyme systems related to the glutamate metabolism, we shall present evidence favoring the idea that the couples glutamate-2-ketoglutarate, hydropyruvate-glycerate, and glyoxylate-glycollate may play important roles in controlling the aerobic or anaerobic destiny of the reducing equivalent produced during glycolysis or in the Krebs cycle.

There is little information in the literature concerning the possibility of using, in mitochondria, for glutamate synthesis, the reducing equivalents produced in the cytoplasm. On the contrary the important work of Slater and co-workers has brought many interesting facts as to the utilization of Krebs cycle intermediates in the reduction of 2-ketoglutarate and ammonia. These authors have produced considerable evidence demonstrating how reducing equivalents utilized in intramitochondrial synthesis could be produced from intermediate(s) of oxidative phosphorylation, which points to the importance, for amino acid synthesis, of energy-rich compounds that can drive the respiratory chain between succinate and NAD.

Following the observation of Krebs and Cohen (1939) that kidney cortex and heart muscle catalyze the dismutation of 2-ketoglutarate and NH_3 to form glutamate, succinate, and CO_2, much work has been devoted to elucidate the mechanism of that reaction. The dismutation has been shown to occur in mitochondria of various sources and to be considerably enhanced by the addition of 3-hydroxybutyrate, malate, isocitrate, and succinate, which demonstrates that 2-ketoglutarate is a relatively inefficient donor of hydrogens for glutamate synthesis.

Slater and co-workers have studied in great detail the requirements for a hydrogen donor for maximum glutamate synthesis. As a result of their extensive studies (see Tager and Slater, 1963a,b; Tager, 1963; Tager *et al.*, 1963; Slater and Tager, 1963), four mechanisms have been proposed for the synthesis of glutamate in the mitochondria.

The Krebs-Cohen Dismutation

The Krebs-Cohen dismutation is the sum of two reactions:

$$\text{2-Ketoglutarate} + NAD^+ \rightarrow \text{succinate} + CO_2 + NADH + H^+ \qquad (7\text{-}1)$$

$$\text{2-Ketoglutarate} + NADH + NH_4^+ \rightarrow \text{glutamate} + NAD^+ \qquad (7\text{-}2)$$

Reaction 7-1 proceeds as follows

$$\text{2-Ketoglutarate} + NAD^+ + CoA \rightarrow \text{succinyl-CoA} + NADH + H^+ + CO_2 \quad (7\text{-}3)$$

$$\text{Succinyl-CoA} + ADP + Pi \rightarrow \text{succinate} + ATP + CoA \qquad (7\text{-}4)$$

indicating that the dismutation must be accompanied by the synthesis of ATP. This has indeed been demonstrated by Hunter and Hixon (1949). Phosphate and phosphate acceptor should be provided for CoA to be regenerated to obtain optimal conditions of the overall sequence.

The dismutation is arsenite sensitive and stimulated by dinitrophenol. As expected, in anaerobiosis glutamate formation is enhanced if a phosphate acceptor is added, and little aspartate is found.

Succinate as Hydrogen Donor

As shown by Tager and Slater (1963b), the hydrogen donor for glutamate synthesis linked with the oxidation of succinate is not the malate formed from succinate. Their results indicate that with succinate as donor, both glutamate and aspartate are found and represent the contribution of succinate and malate to the reduction of 2-ketoglutarate. Addition of a phosphate acceptor has no effect on the amount of glutamate synthesized but stimulates the formation of aspartate. Oligomycin increases the synthesis of glutamate while malonate, amytal, dinitrophenol, and antimycin are inhibitory. The reaction is insensitive to arsenite.

Accordingly the mechanism leading to the reduction of 2-ketoglutarate and ammonia in the presence of succinate must be dependent on a supply of energy and related to succinate as the sole hydrogen donor. The following reaction adequately takes into consideration all the experimental findings:

$$\text{Succinate} + NAD^+ + \sim \rightarrow \text{fumarate} + NADH^+ + H^+ \qquad (7\text{-}5)$$

The reduction of NAD^+ by succinate requires energy (Chance and Hollunger, 1957). This explains the sensitivity of glutamate synthesis to dinitrophenol, amytal, and antimycin, and the stimulating effect of oligomycin. When the aerobic oxidation of succinate is blocked by antimycin, addition of ATP restores glutamate synthesis.

All these results suggest that a high-energy intermediate used for the synthesis of ATP from ADP and inorganic phosphate, may, in certain circumstances, be used to reduce NAD. This phenomenon first described by Chance (1956) and studied by many authors (see, for instance, Snoswell, 1962; Chance, 1961; Klingenberg et al., 1959) is now interpreted as indicating an energy-linked reversal of the respiratory chain (Chance and Hollunger, 1957).

An interesting consequence of this phenomenon is that the synthesis of glutamate brings about an uncoupling of succinate oxidation in a manner ressembling that of dinitrophenol. In the latter case however, energy is dissipated, while in the former the high energy intermediate instead of being used for the synthesis of ATP, provides the necessary energy for the reduction of NAD

$$A \sim I + \text{succinate} + NAD \rightarrow A + I + NADH + \text{fumarate} \qquad (7\text{-}6)$$

Thus in the absence of a phosphate acceptor (or with oligomycin), energy is available for the synthesis of glutamate.

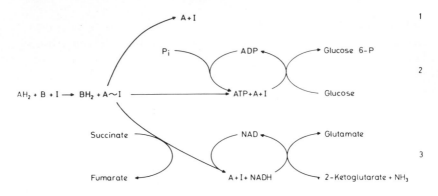

FIG. 7-7. Succinate as hydrogen donor in the synthesis of glutamate. (Slater and Tager, 1963.)

Respiration, as in the case of dinitrophenol, is stimulated. Figure 7-7 is a schematic representation of the mechanism proposed by Slater and Tager (1963) to explain the observed results.

In this scheme an energy-rich intermediate, $A \sim I$ is produced at one of the three phosphorylating sites of the respiratory chain during the aerobic oxidation of the hydrogen donor. It may be used, as in tightly coupled mitochondria, for the synthesis of ATP from ADP and inorganic phosphate (reaction 2, Fig. 7-7). This sequence is inhibited by oligomycin and requires phosphate, ADP, and a phosphate acceptor (such as glucose and hexokinase).

Reaction 1 (Fig. 7-7) leads to the hydrolysis of the energy-rich intermediate with a dissipation of energy. This step is stimulated by dinitrophenol. It explains the stimulation of the respiration observed after application of this compound.

In reaction 3 (Fig. 7-7) the intermediate of oxidative phosphorylation, $A \sim I$, is used to promote the reduction of NAD by succinate. This sequence is inhibited by amytal.

Thus in the absence of phosphate acceptor (or in the presence of oligomycin), 2-ketoglutarate and ammonia provide an alternative pathway for utilization of the high-energy intermediate and lead to a stimulation of respiration.

Malate as Hydrogen Donor

The transamination between glutamate and oxaloacetate is the only known way of synthesizing aspartate. The amount of aspartate found in any system must then be related to the amount of oxaloacetate found. One way of obtaining this compound is through the oxidation of malate:

$$\text{Malate} + \text{NAD}^+ \rightarrow \text{oxaloacetate} + \text{NADH} + \text{H}^+ \qquad (7\text{-}7)$$

As shown by Tager and Slater (1963b) the reduced coenzyme may be used for glutamate synthesis or enter the respiratory chain. In the first instance, glutamate is used in a reaction of transamination for the synthesis of aspartate:

$$\text{NADH} + \text{NH}_4^+ + 2\text{-ketoglutarate} \rightarrow \text{NAD}^+ + \text{glutamate} \qquad (7\text{-}8)$$

$$\text{Glutamate} + \text{oxaloacetate} \rightarrow \text{aspartate} + 2\text{-ketoglutarate} \qquad (7\text{-}9)$$

The sum of reactions 7-7, 7-8, and 7-9 is:

$$\text{Malate} + \text{NH}_4 \rightarrow \text{aspartate} \qquad (7\text{-}10)$$

The sequence is stimulated by arsenite and dinitrophenol sensitive. It is inhibited by antimycin and amytal and the addition of phosphate acceptor, as expected if more NADH produced in Eq. 7-7 enters the respiratory chain. This interpretation has an experimental support in the fact that oligomycin relieves this inhibition. The inhibition by dinitrophenol, antimycin, and amytal finds its interpretation by the consideration of the results showing that the reduction of 2-ketoglutarate and ammonia by malate requires energy provided by the aerobic oxidation of malate.

This energy is not available in the presence of dinitrophenol or on inhibition of the oxidation of NADH by the respiratory chain (antimycin and amytal). The high-energy intermediate must be found at a level situated before the oligomycin sensitive step. It is generated during the aerobic oxidation of succinate or malate. ATP may also serve as energy source in a reaction inhibited by oligomycin. Contrary to what has been found with succinate, there is no stocheiometric relation between the amount of high-energy bond hydrolyzed and the number of molecules of amino acid synthesized. As a result of his study, Tager (1963) proposes the following explanation to account for the fact that energy is required for the synthesis of amino acids coupled to the oxidation of malate: (1) energy is necessary to remove oxaloacetate from malate dehydrogenase which it strongly inhibits; (2) energy is necessary for the transfer of electrons from one pool of NADH (specific for malate dehydrogenase) to another pool specific for the reduction of 2-ketoglutarate and ammonia.

Some workers consider that NADP rather than NAD is the coenzyme specific for glutamate dehydrogenase in mitochondria (Klingenberg and Slenczka, 1959). Accordingly, energy might be necessary for the reduction of $NADP^+$ by NADH as in the following mechanism:

$$NADH + NADP \sim I \rightleftharpoons NADPH + NAD \sim I \qquad (7\text{-}11)$$

Isocitrate as Hydrogen Donor

In the presence of isocitrate, arsenite, oligomycin, antimycin, or amytal have little effect on glutamate synthesis. The same is true if one adds dinitrophenol or a phosphate acceptor to the incubation medium.

In the presence of arsenite the only possible oxidation of isocitrate is that leading to 2-ketoglutarate, therefore the following reaction provides the necessary electrons for the synthesis of glutamate

$$\text{Isocitrate} + NAD^+ \rightarrow \text{2-ketoglutarate} + NADH + H^+ + CO_2 \qquad (7\text{-}12)$$

Since Eq. 7-12 does not require energy, the various inhibitors of the energy metabolism in the mitochondrion should be without effect, as this is indeed found experimentally. If the isocitrate is formed from oxaloacetate, the reaction is inhibited by arsenite.

Glutamate Synthesis as a Control of the Respiratory Chain

In the preceding sections reference has been made to results drawing attention to the fact that crustacea are particularly well suited for the study of the nitrogen metabolism in relation to the activity of the respiratory chain. In the case of a euryhaline species the glutamate synthesis seems to control the respiratory chain in that the reducing equivalents available are either directed toward the synthesis of glutamate or the cytochromes. This could be explained if one postulates that the glutamate dehydrogenase requires NADP as coenzyme in the mitochondria. The transfer of electrons from NAD to NADP would be the step under control:

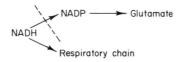

A stenohaline species would lack this type of control.

It is worthwhile noticing that the activity of the dehydrogenases we have studied so far is in all the cases related to the ionic composition of the incubating medium. The interesting observation is that there is an inverse relationship as far as the salt effect is concerned when one considers the glutamate and glycerol-3-phosphate dehydrogenases on the one hand and the dehydrogenases for lactate, isocitrate, and glyoxylate on the other. In the case of a euryhaline species in a concentrated media the activity of the glutamate dehydrogenase is maximal while that of the other dehydrogenases is minimal, thus suggesting that the reducing equivalents produced (in the glycolytic chain for instance) are mainly used for the glutamate synthesis.

If the animal goes into a diluted medium, more of the reducing equivalents available are directed toward the respiratory chain as indicated by the increase in glycerol-3-phosphate dehydrogenase activity and the decrease in glutamate dehydrogenase activity. Thus the synthesis of amino acids should decrease while the oxygen consumption and the ammonia excretion should increase. This is indeed observed with most of the euryhaline species (Florkin and Schoffeniels, 1965; Schoffeniels, 1967a).

This interpretation finds an experimental support in the results obtained by Gilles and Schoffeniels (1965) on the isolated ventral nerve chain of the crayfish. They were able to show that the adjunction of NH_4Cl (10 mM) to the incubating medium of the surviving isolated tissue increases the intracellular pool of free amino acids and decreases the oxygen consumption (from 0.876 down to 0.297 μl O_2 per hour and per milligram fresh weight). This is at variance with the results obtained with the lobster nerve chain where the oxygen consumption is unaffected.

In the case of *Eriocheir sinensis*, another euryhaline crustacean, the situation is a little different since the oxygen consumption is not affected by the transfer of the animal from fresh water to seawater or vice versa (Schwabe, 1933). The amino acid content of the cells however follows the general pattern defined for the euryhaline invertebrates. In this species the glutamate synthesis would seem to act as an uncoupling agent of the oxidative phosphorylation by diverting the energy-rich intermediate from ATP synthesis to the benefit of the synthesis of NADH. Figure 7-8 summarizes this interpretation. In this scheme a constant amount of energy-rich intermediate is produced. Its utilization between ATP or glutamate synthesis is nicely balanced with the consequence that the oxygen consumption is not affected by the pathway followed. In the case of a decrease in glutamate syn-

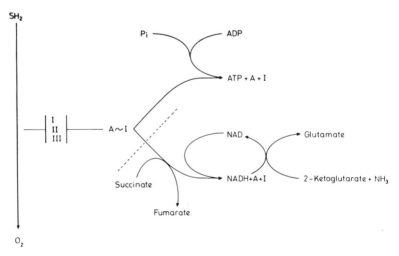

FIG. 7-8. Glutamate synthesis as uncoupler of oxidative phosphorylations. (Schoffeniels, 1968a,b).

thesis, as happens when the animal is in fresh water, more of the energy-rich intermediate could be used for ATP synthesis, that would in turn serve for the active transport of sodium from the dilute medium.

The relationships suggested above, if they satisfactorily explain some of the observations, are by no means considered to be exclusive of other regulatory mechanisms. The proposed schemes are considered as tentative; more information on the intracellular localization of the enzymes involved and on the ionic composition prevailing locally is needed before a more complete picture can be produced. At any rate the comparison with the behavior of the mitochondria extracted from rat liver is already fruitful since in the latter case glutamate synthesis increases together with an enhancement of the oxygen consumption when the synthesis of ATP is prevented by the adjunction of oligomycin (see p. 155).

From what we know of the nitrogen metabolism, the amide glutamine and the amino acids, particularly glutamate, are the most likely candidates in controlling the intracellular pool of ammonia. An interesting relationship stems from the results concerning the effect of inorganic anions on the activity of glutaminase and glutamate dehydrogenase. As is well known, glutaminase activity is enhanced by divalent anion while the converse is true as far as glutamate dehydrogenase is concerned. Thus the ammonia pool that is mainly on the dependence of the activity of these two enzymes could be controlled by the concentration in inorganic anion as indicated in Fig. 7-9. In this schematic representation, the intracellular pool of ammonia is depleted by the activity of glutamate dehydrogenase and that of glutamine synthetase. It is replenished through the deamidation catalyzed by the enzyme glutaminase. In the presence of divalent anions, the intracellular ammonia concentration tends to increase since the production of this molecule is activated while its utilization is inhibited. The reverse is true when the concentration in inorganic anions is low.

Summary

From the above results it may be concluded that euryhalinity

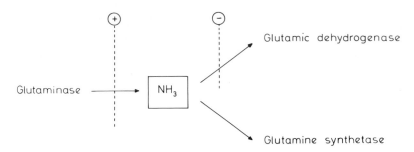

FIG. 7-9.   Control of the intracellular pool of NH_3. Effect of anions on two enzymes: + means activation, − means inhibition. (Schoffeniels, 1968a,b.)

cannot be unequivocally defined as resulting from the control exerted on one single enzyme. It results from the association of a number of adaptations, some morphological, some biochemical, differently distributed among the organisms. This explains why, in many instances, a species very limited as far as its distribution in a given biotope is concerned, may survive experimentally in a medium, the concentration of which is far from that of its natural environment (ecological versus experimental euryhalinity). It is clear that a mechanical device, such as the closing of the opening of the shell in Gastropods, may help a given individual at least temporarily to withstand a dramatic change in the concentration of the outside medium. However, the true euryhalinity can only exist if the animal possesses various mechanisms enabling it to cope with the osmotic stress. We have shown that this implies the possibility to regulate adequately the ionic composition of the blood. The gills and the excretory organs are certainly of prime importance in this respect. On the other hand, it does appear that an efficient control of the intracellular transport of hydrogen is necessary to divert the reducing equivalents toward the glutamate synthesis or the respiratory chain according to the needs imposed by the variation in ionic concentration of the outside medium.

References

Agostoni, A., Vergani, C., and Villa, L. (1966). *Nature* **209**, 1024.
Appella, E., and Tomkins, G. M. (1966). *J. Mol. Biol.* **18**, 77.

Bartels, E. (1962). *Biochim. Biophys. Acta*, **63**, 365.

Bartels, E., Dettbarn, W., Higman, H. B., and Rosenberg, P. (1960). *Biochem. Biophys. Res. Commun.* **2**, 316.

Berl, S., Takagaki, G., Clarke, D. D., and Waelsch, H. (1962). *J. Biol. Chem.* **237**, 2570.

Boxer, G. E., and Devlin, T. M. (1961). *Science* **134**, 1495.

Chance, B. (1956). *In* "Enzymes: Units of Biological Structure and Function" (O. H. Gaebler, ed.) pp. 447–463. Academic Press, New York.

Chance, B. (1961). *J. Biol. Chem.* **236**, 1544.

Chance, B., and Hollunger, G. (1957). *Federation Proc.* **16**, 163.

Cheng, S. C., and Mela, P. (1966a). *J. Neurochem.* **13**, 281.

Cheng, S. C., and Mela, P. (1966b). *J. Neurochem.* **13**, 289.

Côté, L. J., Cheng, S. C., and Waelsch, H. (1966). *J. Neurochem.* **13**, 271.

Datta, A. G., and Katznelson, H. (1957). *Nature* **179**, 153.

Dawson, D. M., Goodfriend, T. L., and Kaplan, N. O. (1964). *Science* **143**, 929.

Devlin, T. M., and Bedell, B. H. (1960). *J. Biol. Chem.* **235**, 2134.

Dewan, J. G. (1938). *Biochem. J.* **32**, 1378.

Duyckaerts, C., and Schoffeniels, E. (1966). *Arch. Intern. Physiol. Biochim.* **74**, 895.

Eisner, T., Alsop, R., and Ettershank, G. (1964). *Science* **146**, 1058.

Ernster, L., and Navazio, F. (1957). *Biochim. Biophys. Acta* **26**, 408.

von Euler, H., Adler, E., Günther, G., and Das, N. B. (1938). *Z. Physiol. Chem.* **254**, 61.

Felicioli, R. A., Gabrielli, F. and Rossi, C. A. (1967). *Life Sci.* **6**, 133.

Florkin, M. (1959). *In* "The Origin of Life on Earth" (F. Clark and R. L. M. Synge, eds.) p. 503. Pergamon, London.

Florkin, M., Duchâteau-Bosson, Gh., Jeuniaux, Ch., and Schoffeniels, E. (1964). *Arch. Intern. Physiol. Biochim.* **72**, 892.

Florkin, M., and Schoffeniels, E. (1965). *In* "Studies in Comparative Biochemistry" (K. A. Munday, ed.) pp. 6–40. Pergamon, London.

Frieden, C. (1959). *J. Biol. Chem.* **234**, 815.

Gilles, R., and Schoffeniels, E. (1964). *Biochim. Biophys. Acta*, **82**, 525.

Gilles, R., and Schoffeniels, E. (1965). *Arch. Intern. Physiol. Biochim.* **73**, 144.

Gilles, R., and Schoffeniels, E. (1966). *Bull. Soc. Chim. Biol.*, **48**, 397.

Hird, F. J. R., and Marginson, M. A. (1966). *Arch. Biochem. Biophys.* **115**, 247.

Hoyaux, J. (1967). "Euryhalinité et osmorégulation chez quelques Gastéropodes de la zone intertidale." Mémoire de Licence en Sciences Zoologiques, Université de Liège.

Hunter, F. E., and Hixon, W. S. (1949). *J. Biol. Chem.* **181**, 67.

Iwatsubo, M., Lécuyer, B., di Franco, A., and Pantaloni, D. (1966). *Compt. Rend.* **263**, 558.

Jeuniaux Ch., and Florkin, M. (1961). *Arch. Intern. Physiol. Biochim.* **69**, 385.

Kaplan, N. O. (1965). *In* "Evolving Genes and Proteins" (V. Bryson and H. J. Vogel, eds.), p. 243. Academic Press, New York.

Kaplan, N. O., Swartz, M. N., Frech, M. E., and Ciotti, M. M. (1956). *Proc. Nat. Acad. Sci. U.S.* **42**, 481.

Kaplan, N. O., Ciotti, M. M., Hamolsky, M., and Bieber, R. (1960). *Science* **131**, 392.

King, E. (1965). *Comp. Biochem. Physiol.* **15,** 93.

Klingenberg, M., and Pette, D. (1962). *Biochem. Biophys. Res. Commun.* **7,** 430.

Klingenberg, M., and Slenczka, W. (1959). *Biochem. Z.* **331,** 486.

Klingenberg, M., Slenczka, W., and Ritt, E. (1959). *Biochem. Z.* **332,** 47.

Koch, H. (1954). *Colston Papers* **7,** 15.

Koefoed-Johnsen, V., and Ussing, H. H. (1958). *Acta Physiol. Scand.* **42,** 298.

Krebs, H. A., and Cohen, P. P. (1939). *Biochem. J.* **33,** 1895.

Lockwood, A. M. P. (1961). *Comp. Biochem. Physiol.* **2,** 241.

McMillan, P. J., and Mortensen, R. A. (1963). *J. Biol. Chem.* **238,** 91.

Markert, C. L., and Møller, F. (1959). *Proc. Nat. Acad. Sci. U. S.* **45,** 753.

Naruse, H., Cheng, S. C., and Waelsch, H. (1966a). *Exptl. Brain Res.* **1,** 284.

Naruse, H., Cheng, S. C., and Waelsch, H. (1966b). *Exptl. Brain Res.* **1,** 291.

Needham, A. E. (1957). *Physiol. Comp. Oecol.* **4,** 209.

Olson, J. A., and Anfinsen, C. B. (1953). *J. Biol. Chem.* **202,** 841.

Palleroni, N. J., and Doudoroff, M. (1956). *J. Biol. Chem.* **223,** 499.

Péquin, L., and Serfaty, A. (1966). *Comp. Biochem. Physiol.* **18,** 141.

Ramamurthi, R. (1967). *Comp. Biochem. Physiol.* **23,** 599.

Rosenberg, P., and Higman, H. B. (1960). *Biochim. Biophys. Acta,* **45,** 348.

Salthe, S. N. (1965). *Comp. Biochem. Physiol.* **16,** 393.

Schlieper, C. (1929). *Z. Vergleich. Physiol.* **9,** 478.

Schoffeniels, E. (1959). "Les bases physiques et chimiques des potentials bio-électriques chez *Electrophorus electricus* L." Vaillant-Carmanne, Liège.

Schoffeniels, E. (1960). *Arch. Intern. Physiol. Biochim.* **68,** 696.

Schoffeniels, E. (1961). *In* "Proceedings Symposium on Comparative Bioelectro-genesis" (C. Chagas and A. Paes de Carvalho, eds.), pp. 147–165. Elsevier, Amsterdam.

Schoffeniels, E. (1964a). *Life Sci.* **3,** 845.

Schoffeniels, E. (1964b). *In* "Comparative Biochemistry" (M. Florkin and H. S. Mason, eds.) Vol. VII, pp. 137-202. Academic Press, New York.

Schoffeniels, E. (1965a). *Arch. Intern. Physiol. Biochim.* **73,** 73.

Schoffeniels, E. (1965b). *Arch. Intern. Physiol. Biochim.* **73,** 157.

Schoffeniels, E. (1966a). *Arch. Intern. Physiol. Biochim.* **74,** 333.

Schoffeniels, E. (1966b). *Arch. Intern. Physiol. Biochim.* **74,** 665.

Schoffeniels, E. (1967a). "Cellular Aspects of Membrane Permeability" Perga-mon, Oxford.

Schoffeniels, E. (1967b). *In* "Traité de Biochimie générale" (M. Javillier, M. Po-lonovski, M. Florkin, P. Boulanger, M. Lemoigne, J. Roche, R. Wurmser, eds.) Vol. III, p. 361. Masson, Paris.

Schoffeniels, E. (1968a). In "Structure and Function of Nervous Tissue" (G. H. Bourne, ed.), Vol. II. Academic Press, New York.

Schoffeniels, E. (1968b). *Arch. Intern. Physiol. Biochim.* **76,** 319.

Schoffeniels, E. and Gilles, R. (1963). *Life Sci.* **2,** 834.

Schwabe, E. (1933). *Z. Vergleich. Physiol.* **19,** 183.

Slater, E. C., and Tager, J. M. (1963). *Biochim. Biophys. Acta* **77,** 276.

Snoswell, A. M. (1962). *Biochim. Biophys. Acta* **60**, 143.

Strecker, H. J. (1953). *Arch. Biochem. Biophys.* **46**, 128.

Tager, J. M. (1963). *Biochim. Biophys. Acta* **77**, 258.

Tager, J. M., and Slater, E. C. (1963a). *Biochim. Biophys. Acta* **77**, 227.

Tager, J. M., and Slater, E. C. (1963b). *Biochim. Biophys. Acta* **77**, 246.

Tager, J. M., Howland, J. L., Slater, E. C., and Snoswell, A. M. (1963). *Biochim. Biophys. Acta* **77**, 266.

Ussing, H. H. (1949). *Acta Physiol. Scand.* **19**, 43.

Ussing, H. H., and Zerahn, K. (1951). *Acta Physiol. Scand.* **23**, 110.

Weil-Malherbe, H. (1957). *In* "Metabolism of the Nervous System" (D. Richter, ed.) p. 474. Pergamon, London.

Weimberg, R. and Doudoroff, M. (1955). *J. Biol. Chem.* **217**, 607.

Whittam, R. (1961). *In* "Proceedings Symposium on Comparative Bioelectrogenesis" (C. Chagas and A. Paes de Carvalho, eds.) pp. 166–168. Elsevier, Amsterdam.

Whittam, R., and Guinnebault, V. (1960). *Biochim. Biophys. Acta* **45**, 336.

CHAPTER VIII

Metabolic Relations in the Production of the Cocoon by the Silkworm

Animals do not always spend their whole postembryonic life in the same medium or in the same form of relationship with the external world. In a number of cases, the change of ecological relations is accompanied by metamorphosis; i.e., morphological changes in nonreproductive structures of the organism. A tadpole has a tail and no legs, it has gills and no lungs. As soon as it matures, the legs show themselves, the tail and gills waste away. The remodeling has produced adaptively modified anatomical structures, and these changes have appeared before the organism entered the new medium. But the remodeling also includes aspects of the nature of biochemical adaptations to the change of medium. The ureotelism which appears in Amphibia as an adaptive defense toward dehydration by evaporation develops through the increase in all the enzymes involved in the biosynthesis of urea (Fig. 8-1). The hemoglobin of the tadpole is exchanged for a new hemoglobin with a lower oxygen affinity and presenting a Bohr effect. The higher oxygen affinity of the tadpole hemoglobin appears as an adaptation to the hazards of changes in oxygen concentration of the liquid medium, while the appearance of the Bohr effect and of the corresponding Haldane effect can be considered as an adaptation to the discharge of CO_2 in the lungs. Bennett and Frieden (1962) have collected a number of aspects of the biochemical adaptations which accompany the passage of the Amphibia from the tadpole to the adult stage.

164

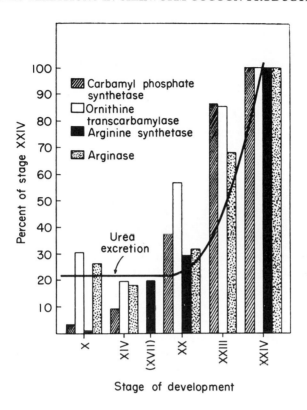

FIG. 8-1. Urea excretion and development of the enzymes of the urea cycle in the metamorphosing tadpole (*Rana catesbiana*). (Brown and Cohen, 1958.)

It is more and more apparent that the metamorphic character-istics are not only of morphological nature but that a number of biochemical adaptations are implied in the exploitation of dif-ferent habitats by different stages of the same organism. These biochemical adaptations subserving ecological relations clearly confirm the fact, emphasized by Snodgrass, that the larva of a holometabolous insect, for instance, cannot be considered as an aspect of embryonic development, nor as an evolutionary reca-pitulation. The adaptations observed during the larval and pupal stages of a butterfly are clearly of ecological nature and certainly more recent than the adult ancestral form. As all the stages of a holometabolous insect have the same genotype, it appears that different sets of sequences of pyrimidine and purine bases are called into play at different times.

A very neat biological clock is at work during the development of *Bombyx mori*. The silk glands, which are modified salivary glands, are brought into increased activity on the fourth day of the fifth instar in the European race. When the cocoon is spun, on the other hand, proteolytic enzymes are liberated in the gland and perform its lysis. During the period separating these two events, the glands accomplish the synthesis of the silk used to spin a cocoon in which the larva wraps itself to go through its pupal life, an aspect of indisputable importance in the wild species of spinning cocoon Lepidoptera.

In the study of insect development, a number of experimental surgical procedures have been devised in order to study the endocrinological aspects of metamorphosis. The methods described here have been developed along another vista, the study of metabolism. Four molts separate the 5 instars of the larval life of *Bombyx mori*. Before each molt, the silkworms cease to eat mulberry leaves, become immobile, and take on a waxy appearance while a triangular spot appears at the posterior portion of the head, under the dorsal prothoracic tegument. At the end of this "triangle" stage, the ecdysis takes place. The fifth instar is the most variable in length, and it becomes necessary to find some good reference points fitted for a metabolic study. From the point of view of feeding, three periods are individualized during the fifth instar. If from the first to the fifth or sixth days, the silkworms are deprived of food, they will die before the nymphal molt takes place. This period, we call the period of indispensable feeding. If the silkworms have received food during this first period, but are fasting from a period lasting from the fifth or sixth day to the tenth or eleventh day, they will nevertheless accomplish their metamorphosis but they will only spin a minimal number of threads, if any. During the fourth instar, a very small proportion of ingested protein nitrogen is directed toward the silk glands, while in the fifth instar, the relative proportions of the nitrogen oriented toward the body tissues and the nitrogen taken by the gland are progressively modified in such a manner that almost all the nitrogen ingested during the second period, the period of facultative feeding, is directed toward the silk glands. This explains why silkworms can dispense with feeding from the fifth or sixth day to the tenth or eleventh day of the fifth instar. If they receive no food during that period, they will, as we have said, only spin a very sparse cocoon, which may amount to only a few threads. If, on the other hand, the silkworms are deprived of their silk gland, an operation we can perform most easily on the third or fourth day after the third larval molt, a number of them will accomplish metamorphosis normally and will reach the adult and reproductive stage.

The last period of the fifth instar is a period of fasting. The emptying of the digestive tract by the release of a brownish fluid gives a good reference point after which the events follow a less variable course than before. This purge or last defecation can be determined precisely if the silkworms, having ceased to feed, are put in paper bags. These are reexamined daily and the day they are marked by the brown spot of the purgation is noted. The liquid of the purge contains a high concentration of phosphates. The silkworm spins in the paper bag, and, four days after the purge, the cocoon is taken out of it. One can then remove one tip of the cocoon, as with a soft-boiled egg, and watch the exuviation corresponding to the nymphal molt.

What happens during the period of fasting with the glandectomized silkworm? This is shown in Fig. 8-2. In both controls and glandectomized larvae, a loss of weight is observed. This is due to evaporation while no water is introduced with food. But the controls spin while the glandectomized do not. If we compare the dry weights of the prenymphs, the difference is also well marked, the dry weight of the glandectomized prenymphs being 1.4 the dry weight of the controls. On the other hand, if controls and glandectomized silkworms are deprived of food at the end of the period of indispensable feeding, the loss of weight will then begin and no difference exists between the glandectomized and the controls. These observations are in agreement with the notion that part of the food taken during the period of facultative feeding is used directly for the formation of silk while another part is set aside in the tissues which, in the controls, are the only source or chemical energy when feeding stops. If the formation of silk is prevented by the ablation of the glands, the weight of the nymphs remains higher.

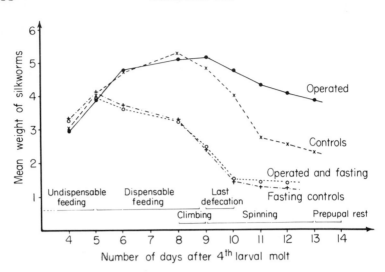

Fig. 8-2.   Variations of weight in normal or glandectomized silkworms, fed or fasting during the period of "facultative feeding." Number of silkworms utilized: 16 glandectomized, 10 controls, 5 fasting glandectomized, 5 fasting controls. (Jeuniaux and Florkin, 1958.)

It appears therefore that the protein metabolism of the silkworm shows an interesting peculiarity: the silk used to spin the cocoon is obtained partly from exogenous nitrogen accumulated directly in the gland reservoir during a specific period of life and partly from the body substance of the larva. The weight of the normal organism is regulated by the accomplishment of this second episode of the accumulation of cocoon material.

There is no doubt that the main contribution of the study of silk to scientific knowledge came from the study of its X-ray diffraction pattern. The concept of the extended polypeptide chain structure of proteins was established as a consequence of studies on silk fibroin. These brilliant studies have somewhat obscured the interest of cocoon spinning as an ecological aspect of protecting the prenymphal stage. Silk is most commonly produced in arthropods by labial glands opening in the head region, but this particular physiological radiation of protein synthesis has many other ecological aspects, for instance in the production of silk by adult insects to protect the eggs in the ootheca, in the production of the spider web, etc.

The term fibroin is generally used to designate the solid fibers of silkworm cocoons, the "cement" of gum coating these fibers is another protein, sericin. As some silk fibers of arthropods, i.e. glossy fibers, have been found in some cases to be made of collagen and others of chitin-protein complex, Rudall proposes to limit the category of fibroins according to their X-ray diagram characteristics and to contrast them with the proteins which, in this respect, are like keratins or like collagens. This classification also fits chemically among fibrous proteins, the fibroin type occurring when there is some 50% of the amino acids as glycine, alanine, and serine, whereas the keratin type occurs when there is considerably less than 50% of the total amino acids made up by the small ones. The collagen configuration, on the other hand, results from a high content of glycine, proline, and hydroxyproline. The physiological and ecological implications of these molecular characteristics are obvious in the study of the physiological radiations of protein synthesis. While the structure of the fibroin of *Bombyx* silk is one of the bases of the definition of the parallel β-type in which the chains are in the extended or β-configuration and lie parallel to the fiber axis, Rudall has discovered that another type, the cross-β, i.e., with polypeptide chains lying across the fiber axis, is found in the so-called silk on the egg-stalk of *Chrysopa flava*. Astbury had already suggested that there should be an α-form of silk and Rudall has already found a number of such forms, so far confined to one group of Hymenoptera, the Aculeata, in bees. These are spun by larvae as is also the case in Lepidoptera. Another interesting adaptation is shown by the cocoon of the gooseberry sawfly *Nematus ribesii* in which the sericigenous glands of the larva are almost completely filled with the fluid precursor of collagen. Now, a very interesting common feature of collagen and fibroin is the fact that both have a high content of the small amino acids, glycine, alanine, and serine. Here we have a mechanism introducing proline in the protein chain, though hydroxyproline has not been found. One of the great interests of the study of fibroins lies in the fact that they provide a transition between the keratine-myosine-epidermine-fibrine (K-M-E-F) group of fibrous proteins, and the collagen group. In fact, by incorporating enough proline in the K-M-E-F fibers, we can, as Rudall (1962) remarks, obtain collagen, while we should as he also remarks, obtain K-M-E-F proteins by incorporating a greater proportion of the larger amino acids in fibroin. Or we may say, by removing large amino acids from K-M-E-F proteins we may get fibroin, as well as by removing proline from collagen.

All these considerations, valuable in biochemical phylogeny, make it important for us to study the metabolic origin of the fibroin of an insect, for instance *Bombyx mori*. To elucidate this point, different experimental approaches have been made. It is of importance to determine the correspondance between organic compounds in the hemolymph at definite periods of the fifth instar, and the different segments of the silk thread forming the cocoon. This rule has been accomplished by T. Fukuda and his Japanese collaborators (1955), working with the hybrids of the Chinese and Japanese races of *Bombyx mori*, and by Fukuda and Florkin (1959a) with the European race (Fig. 8-3). As we shall see, alanine is not one of the free amino acids of hemolymph taken in large amounts by the silk gland for the biosynthesis of fibroin. It is nevertheless absorbed in sufficient quantities to mark the protein synthesized by the radioactive carbon introduced with it. In the experiment shown in Fig. 8-3, at different times (from the first to the twelfth day) of the fifth age, silkworms have received 0.5 μCi of L-alanine-^{14}C in the digestive tract. The amino acid solution has been given by way of a catheter formed by a syringe needle cut at the end, and introduced into the mouth. The animals, fed on mulberry leaves, have spun their cocoons. With the help of a revolving drum of a known circumference, provided with a counter of revolutions, the thread of each cocoon has been unwound and divided into ten equal parts. Each segment was degummed (removal of sericin) by repeated boiling in a 0.02 M solution of sodium carbonate, then washed with ethanol, with ether, and dried. The ten segments obtained from a silk thread have been arranged in a horizontal line on a sheet of cardboard, regularly separated by pieces of adhesive tape, in the same order of their succession in the silk thread, the left end of the horizontal line corresponding to the anterior end of the silk thread (produced at the start of the spinning). The hor-

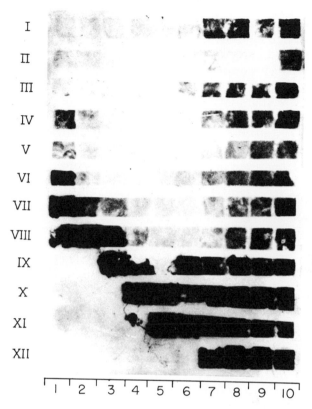

FIG. 8-3. Radioautograph of silk threads from *Bombyx mori*. *Roman numerals* (I-XII) indicate the days following the beginning of the fifth stage on which the L-alanine-^{14}C was introduced into the digestive tract. *Arabic numerals* (1–10) refer to the ten equal divisions of the whole length of each silk thread. (Fukuda and Florkin, 1959a.)

izontal successions obtained in this way, each from one cocoon, have been disposed vertically, each horizontal line corresponding to the day of the fifth age on which the radioactive amino acid had been introduced into the digestive tract, from the first to the twelfth day of the fifth age. An X-ray film was laid on the cardboard and left in contact with it for a period of 18 days, in order to obtain the radioautograph reproduced in Fig. 8-3.

The radioautograph shows three interesting aspects: the high radioactivity of the first segment of silk threads VI, VII, and VIII, the lack of radioactivity of segments 1 and 2 of the silk thread IX, 1−3 of X, 1−4 of XI and1−6 of XII, and the definite increase of radioactivity of segments 7−10, compared to 1−6 in the fibroin of all silk threads. The fibroin is biosynthetized by the cells of the posterior segment of the gland and the introduction of the marked amino acid in the digestive tract has the effect of temporarily increasing the specific activity of ^{14}C in the hemolymph. That this effect is rapidly produced is shown by the observation that one hour after the introduction of an amino acid marked with ^{14}C, the posterior segment of the gland shows a high radioactivity.

We may conclude that the appearance of a high radioactivity in one segment of the fibroin of the silk thread is the result of the simultaneous existence of an active removal of carbon compounds from the hemolymph to the gland and of the high specific activity of ^{14}C in the hemolymph as a consequence of the ingestion of radioactive amino acid. In each of the silkworms used in the experiment it may be postulated that the period of high specific activity of ^{14}C in the hemolymph was of about the same duration. The high radioactivity of segment 1 of thread VI indicates that the posterior part of the gland has taken up a high amount of carbon compounds from the hemolymph a phenomenon appearing early (third day) in the hybrid of the Chinese and Japanese races, in which the duration of the fifth age is shorter (8 days).

The relatively higher radioactivity in the last four segments of the different threads indicates that in all cases an increase in specific activity of ^{14}C has taken place in the hemolymph at the time of the formation of segment 7 of the fibroin thread.

The lack of radioactivity in segments 1 and 2 of IX, 1−3 of X, 1 −4 of XI, and 1−6 of XII indicates that when the radioactive meals have been given respectively on the ninth, tenth, eleventh or twelfth day of the fifth age, the increase of specific activity of ^{14}C in the hemolymph has had no repercussion on the above mentioned segments of these threads, the fibroin of which was already stored in the intermediate portion of the gland (reservoir). The radioautograph of XII indicates that the increase of specific activity of ^{14}C on the twelfth day does not influence the radioactivity of segments 1—6, but does influence the radioactivity of segments 7—10. In spite of the twelfth day being the day on which the animal has stopped feeding, the posterior section of the gland, during the days of spinning, continues to transfer from the hemolymph the carbon compounds used to make the silk and segments 7—10 appear to be synthesized during the days following the twelfth day, i.e., after the cessation of feeding. The relative increase in radioactivity beginning with segment 7 in all threads indicates that the carbon compounds taken by the gland during that period are introduced from the tissues to the hemolymph.

If the first segments of threads IX—XII indicate that the fibroin already stored in the middle segment of the gland is not influenced by the increase of specific activity of ^{14}C produced in the hemolymph, VI—VIII on the other hand, indicate that the radioactive fibroin produced in a given period, as a result of an increase of specific activity of ^{14}C in the hemolymph keeps, in the silk thread, the chronological mark of this high specific activity, in contrast with the fibroin synthesized later, when the specific activity of ^{14}C has decreased in the hemolymph.

From these radioautographic studies it was concluded that, in the European race as well as in the Chinese and Japanese hybrids, the origin of the carbon atoms of each segment of the fibroin of the silk thread can be traced back to the carbon compounds present in the hemolymph at definite periods of the fifth instar, and this correlation has been made precise. These observations suggested that the fibroinogen stored in the reservoir of the gland before spinning is disposed in an orderly fashion corresponding more or less to the moment of its biosynthesis by the silk gland. Fukuda and Florkin (1959b) confirmed this by showing with ^{14}C-glycine that the successive portions of fibroinogen are stored in the reservoir in the order of their synthesis (Fig. 8-4 and 8-5). If this is the origin of the ordered marking of the thread, the train of macromolecules of fibroinogen must move ahead in the reservoir from one end to the other while remaining in order. This was shown by radioautographs of longitudinal as well as transversal sections also studied with respect to the concentration of ^{14}C in the slice, at different times after the introduction of

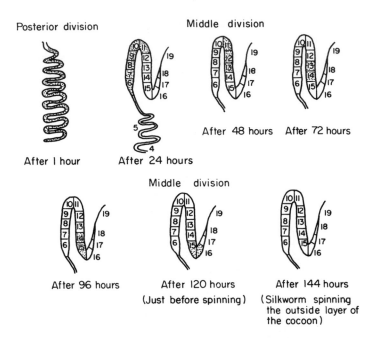

FIG. 8-6. Situation in the silk gland, at different times after the beginning of sixth day of the fifth instar (120th hour after the beginning of the fifth instar) of the fibrinogen synthetized in the posterior division at that time. (Fukuda and Florkin, 1959c.)

mine if phenylalanine can be converted into tyrosine during the fifth instar. That this is the case is shown in experiments in which L-phenylalanine marked on the C-1 has been injected to silkworms. Labeled tyrosine has been carefully isolated from fibroin and its purity well demonstrated (Bricteux-Grégoire *et al.* 1959a). Fukuda has also obtained marked tyrosine by injecting phenylalanine marked on C-2. The same author's claim that alanine can derive from phenylalanine is contradicted by our careful analysis showing that no activity appears in alanine, glycine, or serine.

Alanine, as we have seen, is not significantly taken up by the silk gland contrary to what is found with glycine and serine. It nevertheless is one of the most important amino acids of fibroin,

TABLE 8-1

Free Amino Acids in the Hemolymph of the *Bombyx mori* L.[a]

Days after fourth molt	Condition[b]	Gly[c]	Ala[c]	Ser[d]	Glu (tot)[e]	Asp (tot)[e]	Thr[f]	His[g]	Met[g]	Pro[h]	Arg[i]	Lys[i]	Leu[i]	Ileu[i]	Val[i]	Phe[i]	Tyr[j]
Six	C	57	32	80	213	66	54	104	23	18	45	74	23	14	27	13	—
	O	244	31	102	297	80	136	95	17	48	60	84	28	31	42	17	—
Seven	C	85	42	—	169	—	37	136	15	17	30	83	26	67	29	16	—
	O	272	41	—	255	—	185	123	10	55	51	76	30	32	46	23	—
Eight	C	—	—	66	—	82	—	—	—	—	—	—	—	—	—	—	—
	O	—	—	113	—	83	—	—	—	—	—	—	—	—	—	—	—
Nine	C	99	39	—	244	—	62	203	26	19	43	129	32	27	30	20	—
	O	384	37	—	333	—	227	138	11	71	40	105	21	24	39	25	—
Ten	C	—	—	—	—	—	—	—	—	—	—	—	—	—	—	—	3.5
	O	—	—	—	—	—	—	—	—	—	—	—	—	—	—	—	17.5
Eleven	C	164	35	60	101	47	33	181	49	18	28	156	17	26	14	12	—
	O	620	41	112	276	65	157	112	35	44	16	69	12	20	14	10	—

[a] The values are given in mg per 100 ml of hydrolyzed plasma.
[b] C, control; O, operated (glandectomized).
[c] Duchâteau et al., 1959.
[d] Bricteux-Grégoire et al., 1959d.
[e] Bricteux-Grégoire et al., 1959c.
[f] Duchâteau-Bosson et al., 1960a.
[g] Duchâteau-Bosson et al., 1960b.
[h] Duchâteau-Bosson et al., 1961a.
[i] Duchâteau-Bosson et al., 1961b.
[j] Duchâteau-Bosson et al., 1962.

in which glycine forms 43% and alanine 35%, while serine accounts for 16%. Alanine, as well as glycine and serine, can be derived from other substances taken up by the gland. Glycine can be used to make serine in the rat, pigeon, and certain microorganisms. The metabolic relationship between serine and alanine is also known in mammals. If glycine marked on the carboxyl is given to a silkworm, glycine, serine, and alanine are marked on the C-1 only. If glycine-2-^{14}C is given, a very high specific activity is found in the C-2 of the amino acids glycine, serine, and alanine. Therefore, there exists in the silkworm a direct passage of glycine into serine, probably by the addition of formate. As in the rat, the C-2 of glycine can be a precursor of the C-3 of serine. This comes from the addition of formate. C-1 of glycine, serine, and alanine are also marked by a passage through pyruvate (Bricteux-Grégoire *et al.* 1959 b,c). The silk alanine seems to derive largely from the dicarboxylic acids and their amides, the most abundant constituents of the mulberry leaves (Bricteux-Grégoire *et al.*, 1960).

That the pyruvate is directly used by transamination by the silk gland to form alanine is shown by the observations of Fukuda (1957a, b) and Koide *et al.* (1955). Bricteux-Grégoire *et al.* (1959d) using ^{14}C sodium pyruvate marked in different positions have shown that when pyruvate-1-^{14}C is given per os to silkworms, ^{14}C appears in C-1 in alanine, serine, and glycine of the fibroin isolated from the cocoon. Glucose is also a precursor of amino acids and when labeled glucose is used, a great part of the injected activity is found in the alanine, serine, glycine, glutamic acid, and aspartic acid (Bricteux-Grégoire and Florkin, 1962). It appears therefore that the carbon skeleton of the main amino acids of silk (glycine, alanine, and serine) can derive from any pyruvate precursor but physiologically the glutamic and aspartic acids and their amides appear to provide nitrogen and pyruvate as well, for the synthesis of the amino acids of fibroin. A part of glycine and of serine can come directly form the hemolymph but this is insignificant in the case of alanine. Tyrosine derives from exogenous phenylalanine only (Bricteux-Grégoire *et al.*, 1959a). These results concord with the balance sheet established during the growth of abundantly fed silkworms receiving large amounts of proteins (Fukuda *et al.*, 1961).

In the course of studies on glandectomized silkworms an increase in concentration had been observed after removing the silk gland, not only of glycine, serine, aspartic and glutamic acids, and tyrosine, but also of proline and threonine (Table 8-1). Genetic studies on *Bombyx mori* have not revealed so far the existence of any mutants presenting an abnormal composition of fibroin. A mutant is known (race Nd) producing practically no fibroin (0.6% of the weight of the cocoon according to Watanabe (1959), but synthesizing normal amounts of sericin of normal composition. The cocoons spun by the silkworms of race Nd are consequently lighter and more transparent than those of normal silkworms.

As we have said above, fibroin is secreted by the distal segment of the silk gland, while sericin is secreted by the walls of the reservoir forming the middle part of the gland. In race Nd, it appears that the distal segment is unable to subtract from the hemolymph the constituents used in fibroin synthesis. In silkworms of race Nd, a higher than normal concentration of the fibroinogenous amino acids is observed, contrary to what is found with threonine and proline—this result is in agreement with the fact that these two amino acids are taken up by the gland at the level of the middle part (reservoir) (Brictcux-Grégoire *et al.*, 1964).

The cocoon formation in the silkworm appears as a secretion of a protective sheet for the prenymphal molt and the nymphal life. The protein fibroin, the function of which is related to its physical properties, largely derives from a great excess of specific food rich in glutamic acid, glutamine, aspartic acid, and asparagine. This very specific food is ingested in great excess during the period of activity of the silk gland and thus provides it with a series of amino acids selectively absorbed. The peculiar structure of fibroin, with its recurring sections made up of small amino acids, depends, according to present theories, from a genetic code liberating messenger RNA, specific of the distal part.

The period of activity of the silk gland is accompanied by a derivation of the pyruvate producing constituents of the food toward the posterior part of the gland in which pyruvate is used to synthesize alanine, glycine, and serine, according to pathways corresponding to those found in other forms of animal life. Some glycine and serine can also be included directly in the fabric of fibroin, and this is also the case for tyrosine, taken from the hemolymph. The aspartic and glutamic acids and their amides, forming an important proportion of the proteins of the mulberry leaves, are among the important pyruvate producers in the silk gland. The final section of the silk thread is of endogenous origin. Its production is not only part of the accomplishment of an adaptation subserving ecological functions, but it also subserves the physiological function of regulating the normal size of the nymph by a derivation of a part of the substance of the larva toward the production of silk.

At the time of hatching, another adaptation sets in. It depends on the chemical structure of sericin, which is the specific substrate of an enzyme secreted by the moth and by which sericin is specifically dissolved, permitting the separation of fibroin threads and the passage of the imago (Kafatos and Williams, 1964).

The data reported in this chapter result from a very detailed study of the adaptation consisting of the spinning of a cocoon by the silkworm. The physical properties of fibroin underline this adaptation and depend on the biosynthesis of this protein in the posterior part of the silk gland. It appears, at least in the context of contemporary theories, that the particular nature of a messenger RNA liberated at the level of the cells of the posterior section is responsible for the special differentiation of their cells. This seems to account for the adaptation concerned. But the analysis performed above shows that if the explanation of the *nature* of fibroin is accounted for by a special form of protein synthesis, resulting from a special gene structure, the *existence* of the cocoon results from many other factors: the nature of the food, the behavior which brings the silkworm to this selected food, brings the silkworm to chew it and to swallow it, and the behavior responsible for the enormous food consumption during a part of the fifth instar, the selective extraction of amino acids from the hemolymph into the posterior section of the gland, the regulation of the weight of the nymph, etc. All these factors control the very existence of the cocoon and confirm the polygenic characteristic of the adaptation considered. This could only be shown by a molecular approach, with the condition that the starting point of the enquiry is situated at the level of the organism.

References

Bennett, T. P., and Frieden, E. (1962). *In* "Comparative Biochemistry" (M. Florkin and H. S. Mason, eds.) Vol. IV, pp. 483–556. Academic Press, New York.

Bricteux-Grégoire, S., and Florkin, M. (1962). *Arch. Intern. Physiol. Biochim.* **70**, 711.

Bricteux-Grégoire, S., Dewandre, A., Florkin, M., and Verly, W. G. (1959a). *Arch. Intern. Physiol. Biochim.* **67**, 687.

Bricteux-Grégoire, S., Dewandre, A., Florkin, M., and Verly, W. G. (1959b). *Arch. Intern. Physiol. Biochim.* **67**, 693.

Bricteux-Grégoire, S., Duchâteau, Gh., Florkin, M., and Jeuniaux, Ch. (1959c). *Arch. Intern. Physiol. Biochim.* **67**, 586.

Bricteux-Grégoire, S., Florkin, M., and Jeuniaux, Ch. (1959d). *Arch. Intern. Physiol. Biochim.* **67**, 182.

Bricteux-Grégoire, S., Dewandre, A., and Florkin, M. (1960). *Biochem. Z.* **333**, 370.

Bricteux-Grégoire, S., Duchâteau-Bosson, Gh., Jeuniaux, Ch., and Florkin, M. (1964). *Arch. Intern. Physiol. Biochim.* **72**, 489.

Brown, G. W., and Cohen, P. P. (1958). *In* "A Symposium on the Chemical Basis of Development" (W. D. McElroy and B. Glass, eds.) pp. 495-513. John Hopkins, Baltimore.

Duchâteau, Gh., Florkin, M., and Jeuniaux, Ch. (1959). *Arch. Intern. Physiol. Biochim.* **67**, 173.

Duchâteau-Bosson, Gh., Bricteux-Grégoire, S., Florkin, M., and Jeuniaux, Ch. (1960a). *Arch. Intern. Physiol. Biochim.* **68**, 275.

Duchâteau-Bosson, Gh., Florkin, M., and Jeuniaux, Ch. (1960b). *Arch. Intern. Physiol. Biochim.* **68**, 327.

Duchâteau-Bosson, Gh., Jeuniaux, Ch., and Florkin, M. (1961a). *Arch. Intern. Physiol. Biochim.* **69**, 369.

Duchâteau-Bosson, Gh., Jeuniaux, Ch., and Florkin, M. (1961b). *Arch. Intern. Physiol. Biochim.* **69**, 485.

Duchâteau-Bosson, Gh., Jeuniaux, Ch., and Florkin, M. (1962). *Arch. Intern. Physiol. Biochim.* **70**, 287.

Fukuda, T. (1957a). *J. Biochem. Tokyo* **44**, 505.

Fukuda, T. (1957b). *Nature* **180**, 245.

Fukuda, T., and Florkin, M. (1959a). *Arch. Intern. Physiol. Biochim.* **67**, 185.

Fukuda, T., and Florkin, M. (1959b). *Arch. Intern. Physiol. Biochim.* **67**, 190.

Fukuda, T., and Florkin, M. (1959c). *Arch. Intern. Physiol. Biochim.* **67**, 214.

Fukuda, T., Kirimura, J., Matuda, M., and Suzuki, T. (1955). *J. Biochem. Tokyo* **42**, 341.

Fukuda, T., Duchâteau-Bosson, Gh., and Florkin, M. (1961). *Arch. Intern. Physiol. Biochim.* **69**, 701.

Jeuniaux, Ch., and Florkin, M. (1958). *Arch. Intern. Physiol. Biochim.* **66**, 552.

Kafatos, F. C., and Williams, C. M. (1964). *Science* **146**, 538.

Koide, F., Nagayama, H., and Shimura, K. (1955). *Nippon Nogei-Kagaku Kaishi* **29**, 987.

Rudall, K. M. (1962). *In* "Comparative Biochemistry" (M. Florkin and H. S. Mason, eds.) Vol. IV, pp. 397 – 433. Academic Press, New York.

Watanabe, T. (1959). *J. Sericult. Sci. Japan* **28**, 251.

CHAPTER IX

Concluding Remarks

The essential object of biology is situated *sub specie evolutionis*. The recognition of the biochemical unity of cells has brought to the theory of evolution its most decisive proof. On this unity, the diversity of organisms is superimposed. Since Lamarck we have learned to consider the diversity of species as the result of adaptations to different environments. Adaptation as understood here, is an organismic concept. It is however possible to individualize biochemical peculiarities within the more complex picture of all the adaptive features. A proper approach to the study of adaptation must start from the consideration of the relation organism-environment at the level of the community or of the organism, and progressively proceed from this organismic starting point to the underlying molecular aspects. Many examples have been presented in the preceding pages showing that the adaptation to a given environment cannot be traced down to a single molecular mechanism. Mechanical device as well as morphological feature appear together with some peculiar properties of certain molecules as representing the array of characteristics explaining why an animal succeeds in a given environment. All these factors are not equally important and their hierarchy has to be established for the phenomenon to be fully understood. To be convincing, such a study must be very detailed. This was accomplished in Chapters V, VI, and VII in the case of the adaptation of euryhaline invertebrates to media more diluted than seawater. We have shown that in certain of these euryhaline forms, this adaptation results from the effects of an anisosmotic regulation of the body fluids, kept at a higher concentration than the fluid medium, and of an isosmotic intracellular regulation bringing the cells into osmotic equilibrium with the more or less diluted body fluids.

The regulation of the extracellular fluid composition results from several biochemical adaptations at the level of the organs particularly well suited to exchange matter with the outside medium. The gills, the digestive tract, and the excretory organs have evolved special mechanisms that render possible a control over the composition of the *milieu intérieur* of the organism. Isosmotic intracellular regulation results also from the emergence of new ways of controlling the activity of key enzymes related to the intracellular transport of hydrogen. Thus the amino acid synthesis, and more specifically that of glutamate, appears as a means of controlling the respiratory chain.

Isosmotic intracellular regulation, thermogenesis in homeotherms, flight in insects, and adaptation of diapausing insects to very low temperature appear as physiological radiations of the various biochemical systems that are responsible for the transfer of reducing equivalents. By acquiring a differential control over intracellular hydrogen pathways the cell may synthesize amino acids, glycerol, or produce calories, thus permitting the colonization of the appropriate medium.

In all these cases part of the biological adaptation rests on a peculiarity of some protein molecules, and consequently is in direct relationship with the genotype. It is certainly true that most of the control mechanisms are allosteric in nature and too little is known about the factors controlling their appearance to bring the consideration of this adaptation in relation to a single operon. The morphological adaptation results on the other hand from modifications not easy to identify inside a long multidimensional network of interrelations with numerous genes. This could be the case for the allosteric sites, if it would appear that their emergence is controlled by cytoplasmic factors.

Another adaptation has been studied in great detail in Chapter VIII. It deals with the spinning of a cocoon by the silkworm. The physical properties of fibroin explains the protective function of the cocoon. But a detailed analysis shows that the existence of the cocoon results from many other factors, thus confirming the polygenic characteristics of the adaptation.

In Chapters III and IV, biochemical traits that are adaptive aspects resulting from the properties of macromolecular units have been enumerated. With the apparition of molecules having a property of repellency for some species or active in the regulation of courtship or other social activities, the adaptive nature of the synthesis leading to the molecule in point can hardly be questioned. In many cases, however, the biosynthesis of new molecules cannot be described as adaptation. In the case of a food attractant it is clear that the synthesis of the attractant cannot be described as an adaptation of the plant but rather that the predator is the one adapted to respond to the information carried by the molecule.

To summarize briefly, when dealing with the problem of adaptation it is certainly ill-advised to dissociate the organismic and the molecular aspects. It is true that some function of an organism may be traced down to specific properties of a given molecule. But few adaptations can be related to the properties of one single molecule, thus indicating the frequent polygenic nature of the phenomenon. It remains certain, however, that we owe to what is generally called molecular biology the knowledge enabling us to define the concept of homology at the molecular level. This gives biologists the possibility to seize the concepts of adaptation, no longer in their usual phenomenological context, but at the physicochemical scale of dimensions, thus permitting the unraveling of the molecular evolution that holds the key to morphological and physiological adaptations.

Author Index

Numbers in italics refer to the pages on which the complete references are listed.

A

Abbott, B. C., 70, *73*
Adelstein, S. J., 55, *73*
Adler, E., 139, *161*
Adler, J., 22, *39*
Agostoni, A., 141, *160*
Allee, W. C., 104, *110*
Allen, M. B., 47, 65, *72*
Allison, J. B., 24, *39*
Alsop, R., 112, *161*
Andrewartha, H. G., 57, *72*
Anfinsen, C. B., 64, 72, 139, *162*
Apella, E., 132, *160*
Arrhenius, S., 46, *72*
Avi-dor, Y., 30, *41*

B

Backus, M. P., 31, *39*
Bailey, A. J., 62, 63, *72*
Baker, J. E., 24, *42*
Barbier, M., 37, *39*
Bar-Geev, N., 30, *41*
Baroody, A. M., 38, *42*
Bartels, E., 119, *161*
Barth, R. H., 37, *40*
Batham, E., 30, *40*
Bazinet, M., 36, 39, *44*
Beach, E. F., 82, *87*
Beadle, L. C., 105, *109*
Beatty, D. D., 70, 72, *74*
Beck, S. D., 25, *40*
Becker, G., 33, *40*
Beckmann, R., 36, *40*
Bedell, B. H., 150, *161*
Bennett, T. P., 164, *181*
Bergeim, O., 80, 82, *88*
Bergmann, E. D., 38, *41*
Bergström, S., 37, *40*
Berl, S., 124, *161*
Beroza, M., 36, *42*
Bieber, R., 141, *162*

186

Binyon, J., 105, *109*
Bird, A. F., 27, *40*
Bishop, H., 31, *40*
Black, E. D., 36, 39, *44*
Block, R. J., 82, *87*
Blum, M. S., 25, 33, *40*
Boden, B. P., 70, *73*
Bonner, J. T., 32, *40*
Bossert, W. H., 35, 39, *40*, *44*
Boxer, G. E., 150, *161*
Boylan, D. B., 33, *40*
Brattsbröm, H., 105, *109*
Bricteux-Grégoire, S., 59, 72, 78, 84, 87, 88, 98, 99, 101, 104, 105, 106, *109*, *110*, 176, 177, 178, 179, *181*
Bridges, C. D. B., 68, 69, 70, *73*
Brown, G. W., 165, *181*
Brown, L. A., 50, *73*
Brown, P. K., 68, *74*
Bullock, T. H., 49, 50, *73*, *74*
Burkholder, W. E., 36, *43*
Burlington, R. F., 54, 55, *73*
Burton, A. C., 91, *111*
Butenandt, A., 33, 36, *40*
Butler, C. G., 35, 37, *40*

C

Cahn, R. D., 54, *73*
Callow, R. K., 37, *40*
Calvin, M., 19, *40*
Camien, M. N., 77, 78, 79, 81, 87, 88, 103, 106, *109*
Campbell, L. L., 65, 66, *73*
Carlisle, D.B., 28, *40*
Carr, W. J., 37, *40*
Caul, W. F., 37, *40*
Cavill, G. W. K., 33, *40*
Chadha, M. S., 33, *40*
Chance, B., 153, *161*
Cheng, S. C., 125, *161*, *162*
Chidester, J. C., 33, *40*

Genus and Species Index

A

Acanthomyops, 36
Acanthomyops claviger, 33
Acetobacter melagenum, 126
Achromobacter ichthyodermis, 47
Achylia, 31, 32
Acmaera scutum, 93
Aeshna, 85, 86
Alligator mississipiensis, 47
Allolobophora, 71
Allomyces, 32
Ancylostomum caninum, 47
Anodonta cygnea, 78, 81
Apis mellifica, 85
Aplysia, 26
Aplysia juliana, 29
Archiulus sabulosus, 25
Arenicola, 72
Arenicola marina, 78, 105
Ascaris, 63
Astacus astacus, 78, 106, 108
Astacus (Austropotamobius) pallipes
    Lereboullet (=Saxatilis Bell), 106
Astacus fluviatilis, 81, 121, 133,
    142-148
Asterias rubens, 78, 105, 106
Atta rubropilosa, 33
Attagnus megatoma, 36
Avena, 71

B

Bacillus coagulans, 66
Bacillus stearothermophilus, 65, 66
Bacillus subtilis, 47
Balanus nauplii, 71, 72
Balanus perforatus, 71
Birgus latro, 104
Bombus terrestris, 36
Bombyx mori, 28, 36, 58, 84, 166,
    169-171, 177, 179
Brachinus, 25

Brachiorus, 35
Bracon cephi, 58
Branchiostoma (Amphioxus lanceola-
    tum), 14
Buccinum undatum, 82, 83
Bufo, 33, 72
Bufo bufo, 33, 47
Bufo calamita, 33
Bupalus piniarius, 31

C

Callinectes, 149
Cancer pagurus, 142, 144
Carabus auratus, 25, 26
Carassius auratus, 51
Carausius, 85
Cardisoma armatum, 104
Carcinus moenas, 78, 97, 99, 100, 103,
    104, 106, 115, 118, 121, 140, 144,
    149, 150
Cassis tuberosa, 27
Chrysopa flava, 169
Chydorus, 72
Cimex lectularis, 47
Citellus tridecemlineatus, 54
Clostridium botulinum, 47
Clymenella mucosa, 48
Clymenella torquata, 48
Coenobita perlatus, 104
Corynebacterium diphtheriae, 47
Creophilus, 24
Cryptocercus punctulatus, 31
Cyclops, 72
Cyprinus, 69
Cyprinus carpio, 148

D

Daphnia, 50, 51
Deilephila euphorbiae, 59
Dendroctonus varivestis, 47
Diadema antillarum, 27
Diaptomus, 72

193

Subject Index

A

196